Understanding the Eurovision Song Contest in Multicultural Australia

Jessica Carniel

Understanding the Eurovision Song Contest in Multicultural Australia

We Got Love

palgrave
macmillan

Jessica Carniel
School of Arts and Communication
University of Southern Queensland
Toowoomba, QLD, Australia

ISBN 978-3-030-02314-0 ISBN 978-3-030-02315-7 (eBook)
https://doi.org/10.1007/978-3-030-02315-7

Library of Congress Control Number: 2018957681

This Palgrave Pivot imprint is published by the registered company Springer Nature Switzerland AG
The registered company address is: Gewerbestrasse 11, 6330 Cham, Switzerland

Contents

CHAPTER 1

Introduction

Abstract The Eurovision Song Contest has screened in Australia since 1983, attracting more viewers each year, particularly since Australia was invited to become a participating nation in the contest in 2015. This chapter contextualises the song contest's popularity in Australia within global fan studies, including a reflection upon the author's own "aca-fan" status. Eurovision fans in Australia and around the world value the contest for its cosmopolitanism, and its ethos of diversity and inclusion. The chapter introduces the book's argument that although Australian Eurovision fandom is indeed a result of Australian multiculturalism, the dominant narrative focused on Australia's connection to Europe via migration is overstated and overlooks the complex diversity of Australian society in a globalised world.

Keywords Cosmopolitanism · Fan studies · Diversity · Inclusion · Multiculturalism · Australia

My alarm goes off at 4.50 a.m. Groggily, I rise and dress myself warmly against the late autumn chill. As the kettle boils, I turn on both the television and my laptop. "Good morning, Australia!" I tweet, punctuating my greeting with emojis of a steaming cup and a croissant, and the hashtags #Eurovision and #SBSEurovision. I receive a handful of likes that let

J. Carniel, *Understanding the Eurovision Song Contest in Multicultural Australia*,
https://doi.org/10.1007/978-3-030-02315-7_1

me know others around Australia and the rest of the world are online. As I settle under my nest of blankets on the couch, tea in hand, the cat jostling with computer for space on my lap, the strains of "*Te Deum*" float from my television.

For the past four years, this has become the annual ritual of the Australian fan of the Eurovision Song Contest (ESC). Prior to 2015, when Australia transitioned from mere audience to competing participant in the contest, the Australian telecast of the events, from the semi-finals to the grand final, were broadcast on delay so that viewers watched during a more convenient evening time slot. Now, as viewers of a participant country, Australians are eligible to vote in their semi-final round and the grand final. As the time difference between Australia and most (western and central) European zones is between seven and nine hours, the live telecast of Eurovision occurs at 5 a.m. Australian Eastern Standard Time; viewers on the west coast of Australia must rise at 3 a.m.

Over half a million Australian viewers made this early start in 2017 alone, with twice as many viewing the delayed broadcast in the evening—although it must be acknowledged that many of these evening viewers will be watching it for the second time that day (Knox 2017). Viewer numbers have decreased after a peak of 4.2 million total for all three finals shows in 2015, when Australians were dazzled—and perhaps bemused—by the novelty of being invited to participate in Eurovision for the first—and, we believed then, the only—time in the contest's history. In 2018, the total viewership for all three finals, including both morning and evening broadcasts, was just over one million. The decrease in numbers should not be read as a decline in the contest's popularity in Australia. Rather, the difference is the shedding of the casual viewers to reveal the stalwart, dedicated audience and, at the core of this, the true fans who are dedicated to all things Eurovision more than that one week in May.

This is the story of those fans and the broader audience of the Eurovision Song Contest in Australia, the very people who have to contend with the question I have sought to answer in this book: why do Australians love the Eurovision Song Contest? After all, many do not hesitate to point out, it is a *European* song contest and Australia, despite its colonial history and its mass European migrations, particularly after the Second World War, is not part of Europe. Isn't it in the Asia-Pacific? Protestations became more vehement when Australia officially joined the competition. As Graham Norton, the BBC's commentator for Eurovision, famously opined, "I know some countries aren't technically in Europe but, come on—Australia is on the other side of world…I've got nothing against Australia. I just think it is kind of stupid" (quoted in Wooton 2016).

Other criticisms Australian fans must contend with are less geopolitical, and more cultural. While Eurovision's reputation for kitsch and camp spectacle is a draw card for many, for others it is a source of derision. Furthermore, European pop music in general is often seen as inferior to the offerings from the dominant pop music industry in the United States, despite the prominence of Swedish songwriters and producers in many major pop hits over the past twenty years and other notable musical incursions from Europe. Eurovision songs in particular (with some convenient exceptions) are viewed by sceptics to be not the best Europe has to offer but, rather, the worst: misguided novelty acts that ape outdated trends in American pop music. After all, if the music was so good, would it not break into the global market organically? Australia's own pop music scene is vibrant but its global hits are relatively rare, perhaps diluted by the sheer mass of product available in the global music market (Ferreira and Waldfogel 2013). There are, of course, historical exceptions, such as the so-called "princess of pop" Kylie Minogue, who has been Madonna-like in her ability to reinvent herself and carve out a long career (Minogue first entered the scene in the 1980s and released her most recent album in 2018) and pop-rock crossover sensation INXS. More recent successes include artists like Gotye, Troye Sivan, and Iggy Azalea, although the latter's career in the US has been criticised for its use of cultural appropriation (Eberhardt and Freeman 2015). Overwhelmingly the local music scene has been dominated historically by variations of rock music and "indie" or "alternative" stylings, with local hip hop and dance music rising in popularity since the 2000s. Here, pop is defined as music that is accessible, "produced commercially, for profit, as a matter of enterprise not art" (Frith 2011, 94). Pop music as a genre is, in this context, often viewed as something "foreign" that Australians consume as an import rather than something that Australia manufactures itself. Furthermore, in a manifestation of the Australian cultural cringe (Phillips 1950), local pop offerings can be perceived as inferior, lacking the polished production values of the American industry, *unless* they also find success outside of the Australian market. Much European pop, particularly from non-English speaking countries, is also seen in this light, which can in turn affect how Australians approach the Eurovision Song Contest.

This anxiety about being on the margins of global pop music culture mirrors other forms of marginalisation associated with the contest. Eurovision's great popularity with fans from LGBTIQA communities around the world, including Australia, is both a boon and a bane. On the one hand, it offers a strong market ready for new pop music, and in return has become a de facto international pride festival, about which fan and industry perspectives can be ambivalent. On the other hand, this association has been the source of political tensions between conservative and liberal participating states, which has important economic, political, cultural, and social ramifications. Turkey, for example, has not participated in Eurovision since 2014. In addition to concerns about participation costs, the voting system, the inequity of the "Big 5" system, and other significant regional politics (Vuletic 2018; *Times of Israel* 2018), officials from Turkish broadcaster TRT have criticised representations of queerness at the contest as inappropriate for family programming (Reuters 2018). While the EBU have stated that TRT and Turkey would be welcome in the contest again, its statement equally emphasised diversity and inclusivity as core values of the contest. These principles were put into practice in 2018 when the EBU terminated its partnership with Chinese broadcaster MangoTV after it censored, amongst other things, the Irish performance for its depiction of same-sex relationships and any rainbow flags visible in the crowd (EBU 2018).[1]

By participating in Eurovision, Australia is now participating in an international conversation around various economic, political, cultural, and social issues that are articulated in various ways through the administration and production of the Eurovision Song Contest. Its audience is also part of a transnational corpus made up of a range of local and international communities, bound by a shared interest and connected via the multitudinous nodes of social media networks. When I tweet "Good morning, Australia", my use of the #SBSEurovision and #Eurovision hashtags signals my desire to connect to both local and global fans and viewers (acknowledging that the two subject positions may not necessarily align in the personal identification of some individuals). The text of my tweet highlights both my spatial and temporal distance from Europe and proximity to fellow Australians (who, given the sheer magnitude of the

[1]The Albanian performance was also censored because the performers sported visible tattoos.

Australian continent, may also be quite distant but nevertheless bound in the imagined community), but the act of tweeting simultaneously closes down this distance and creates proximity through our shared engagement with the Eurovision Song Contest live broadcast.

Australia has been an associate member of the European Broadcasting Union (EBU) since 1950. Yet the idea that Australian audiences would be interested in the contest at all, let alone rising in the small hours of the morning to engage in multimedia exchanges about the meaning and significance of acts or host commentary was likely unimaginable when the EBU first established the Eurovision Song Contest in 1956 as a means of cultivating a unified European television audience (Bourdon 2007; Gripsrud 2007). Comprising just a single evening of song with just seven contestants, Lys Assia's 1956 winning song for Switzerland, "Refrain (Schweiz)", was a far cry even from the kind of pop music emerging in the 1950s, let alone the "fast food music" of now, to draw upon the criticisms of popular music mounted by the 2017 Portuguese winner, Salvador Sobral. Brian Singleton et al. (2007) characterise the early Eurovision Song Contest as a highly respectable event until its shift towards large, fan-populated arena audiences in the 1990s. It is now a requirement that the host city provide a venue that can accommodate a minimum of 7000 audience members who are an important prop in the presentation of the television spectacle of the modern contest (Hay and Kanafani 2017). By contrast, Lys Assia performed her *chanson* in front of a live orchestra and a relatively small live audience who were seated and dressed in evening attire. Those watching at home—or, rather, those listening to the radio broadcast—were likely in the company of their family, not a living room of friends in outlandish costumes competing to make the wittiest comments to both their immediate and virtual communities.

Technological advances aside, the political and cultural climate in which Eurovision operates has vastly changed. Even prior to the geography-defying inclusion of Australia, the boundaries of the contest have expanded beyond any post-World War Two conception of "Europe" (Fricker and Gluhovic 2013). Eurovision began in the early days of the Cold War (Vuletic 2018), so was focused upon the Western states (and, eventually, Israel) until the explosion of newly independent, post-Soviet nation-states in the 1990s and early 2000s. The forces of globalisation, as much as time, have shifted both musical taste and the role of music, language, and performance in articulating national identity on the international stage of the Eurovision Song Contest. Viewed somewhat cynically, the inclusion of Australia highlights the role of market forces in television as much as it might be interpreted more sentimentally as a show of cultural connection and goodwill.

It is easy to make an objective observation that the Australian involvement with the Eurovision Song Contest is a natural product of these various forces—globalisation, shifting markets, technological change—but it does not quite answer the question of why Australian audiences connected to the song contest to begin with, let alone how it developed millions of Australian viewers and culminated in an invitation to participate. Through conversations with fans, producers, and commentators, I have uncovered a variety of personal reasons, pragmatic reasons, and overarching theories accounting for the popularity of Eurovision with Australian audiences. The dominant narrative in both official and popular rationalisations about Australian Eurovision audiences, perpetuated by the SBS and the EBU, other media, and fans themselves, centres upon Australia's rich multicultural—and specifically European—heritage. This is what I term here the "European connection" narrative. A secondary narrative about gay and queer culture also appears in some fan discussions but plays a far lesser role in that circulated by official channels. What I ultimately uncovered was that multiculturalism and queerness are certainly part of the texture of Australian Eurovision fan culture, but they are not the driving forces for most individual fans. That is, fans do reiterate the dominant narratives when explaining why Australians enjoy Eurovision in a broad sense, but these rarely align with their own reasons. Rather, the top three reasons for loving Eurovision as cited by the Australian fans in my study are fun, the contest's camp or kitsch factor, and, of course, the music. The feeling of belonging to one's cultural or ethnic heritage ranked sixth, equal to its celebration of personal values.

Although a large number of my participants (39%) identified as LGBTIQA, queer community and associations were not a primary factor in Australian Eurovision fandom. Not a single respondent in the survey nominated it as the primary reason they loved watching Eurovision, and only 4.3% nominated it as a secondary reason. As always, statistics only tell part of the story. In the qualitative responses of both the surveys and interviews, many respondents—both straight and LGBTIQA-identifying—referred to elements of queer culture as part of their Eurovision experience. This varied from, for example, citing the importance of Dana International's victory in 1998 in "outing" the contest to general praise for the idea of a global event that is inclusive of queer communities. Further probing did reveal individual stories of romance found at Eurovision, or of belonging found within the fandom because of its acceptance of queer identities. Ideas of LGBTIQA identity, inclusion, culture, and community thus form an important part of the texture of the event and its fan culture, even if it is not the primary connection identified quantitatively, but the narrative was not as dominant as the European connection. The queer dimensions of Eurovision and the importance of the competition to the Australian LGBTIQA audiences remain useful for understanding the dynamics of belonging in Australian society and to the Western values that Eurovision is said to celebrate. Consequently, these are integrated here into a broader conceptualisation of diversity and multiculturalism as they relate to the "European connection", but it is acknowledged that further consideration of this dimension is required and will be the subject of a future study.

Fan stories return constantly to ideas of community in various forms and connection to the concept of unity that is offered—not unproblematically—by this song contest to those that may feel marginalised in other ways. Unique to the Australian context is the way the belonging to the ESC, whether as a fan or as a member of a competing nations, addresses a sense of geographical and cultural isolation that is far more acute than that experienced by the border-states of Europe. This book concentrates predominantly upon the "European connection" narrative. It engages with prevailing themes of multiculturalism and diversity, the desire for belonging to communities at various personal, state, and global levels, and anxieties about Australian isolation from global cultures. It examines how these dimensions are experienced, challenged, and leveraged in the production and consumption of Eurovision in Australia.

The list of fan reasons for watching Eurovision provided above derive from the second of two datasets that comprise this research. The first dataset is taken from fourteen semi-structured interviews conducted with fans over 2016–2017, and additional interviews with key personnel in the production and distribution of Eurovision in Australia. Fan interviewees were recruited via social media (specifically Twitter and Facebook), snowballing, and via a recruitment message included in my by-line for articles I wrote for the SBS Eurovision website in 2016. Representatives from SBS and its Eurovision production partner BlinkTV were approached directly. Interviews were conducted via Skype or telephone in a conversational, semi-structured style that allowed participants to talk at length about their memories and experiences of the Eurovision Song Contest. All interviews were transcribed and participants were able to review their responses; any significant revisions or additions made by participants are referred to here as a follow-up interview. These interviews provide important and often poignant qualitative, narrativised responses to Eurovision in Australia, but the small sample size limited the study's capacity for observations about broader trends and habits. Subsequently, an online survey was developed from several of the key themes and questions that arose from the interviews. This online survey (n=108) was conducted in 2017, using a mixture of multiple choice questions, scales, and open-ended questions that probed further into the underlying rationale for particular responses. The survey was distributed via the same social media networks as the interview recruitment; I am particularly indebted to the support of several well-known figures in Australian Eurovision fan circles, including bloggers, podcasters, fellow academics, and even former commentators, who assisted in the circulation of this survey. All quantitative data reported in this study derives from this survey; it does not include quantification of responses from the qualitative interviews unless otherwise specified. Survey participants were self-selecting and no demographic quotas were applied; accordingly, the statistics reported here are specific to my sample and may not be representative of all Eurovision fans or viewers in Australia. The open-ended questions used in the survey also contribute significantly to the qualitative data cited throughout, but as the survey responses were anonymous, any qualitative responses taken from it are attributed to "fans", "participants", or "respondents." Conversely, interviewees are referred to by their first names or by a first name they selected as their pseudonym, and are referred to collectively as "interviewees" in order to distinguish datasets. A third dataset emerges from participant observation at the 2018 contest in Lisbon, Portugal. This is used to inform the discussion throughout, culminating in the epilogue, but primarily underpins further forthcoming research into Eurovision-related tourism (or, as I term it, "Tourovision") and the experience of the live event.

It is perhaps also important to establish—or perhaps confess—that I am what Henry Jenkins (2006, 4) has termed an "aca-fan": an academic who is studying a fandom to which they belong, or a text of which they are a fan. In mapping out participatory (Jenkins 1992) and convergence cultures (Jenkins 2008) of fandom, aca-fandom has emerged as an additional way by which contemporary fandom problematises the relationship between production and consumption. John Fiske (2002, 46) differentiates fans from "normal" audience members (also referred to by some scholars as passive consumers) by the way in which they read into texts excessively, finding and filling gaps in meaning; aca-fans are excessive readers par excellence. Just like fanfiction writers, they contribute to the body of texts produced by fans about their object of fandom and can come to influence the production of the object itself. Paul Jordan, for example, built a media profile as "Dr Eurovision" that later led to several years working as the contest's Communications Officer, while members of the Australian delegation, from commentators to artists, frequently declare their own Eurovision fandom even as they contribute to its production for Australian audiences.

I initially watched the contest as a child with my family before a sullen teenage hiatus (oddly coinciding with the years during which I was an avid consumer of SBS's cult content, which, as will be discussed in the Chapter 3, played an important role in growing Australia's Eurovision audience). I returned to the contest as an adult, watching consistently since 2002, and for many years co-hosted an annual party with a friend. My own experiences of Eurovision informed my initial interest in the event as an area of research but also meant that my subjective experiences needed to be considered critically in research design and process. For example, I also believed the official narratives about Eurovision and so initially designed the project to target those who identified as either culturally diverse or LGBTIQA because it aligned with my own personal experiences as a second generation Australian who celebrated the contest with a cohort of queer and queer-allied friends. It soon became clear that targeting these two broad groups limited the scope of the research to the point that it was not an accurate reflection of Eurovision fandom in Australia. Some effects were more positive. My own fan status was useful for connecting to participants in a peer-to-peer manner that facilitated open and dynamic conversations in the interviews; interviewees were often stated that they were not only pleased that the research was being conducted, but that they enjoyed the opportunity to talk about Eurovision openly and without judgment. Being an aca-fan can also blur the line between participant and observer in fan cultures, particularly within the online space. Much of Australian Eurovision fandom has moved into online spaces as a strategy to address the challenges of geography and time difference, but it is also a space where Eurovision scholars in Australia and internationally engage in direct commentary and discussion with each other and fans simultaneously.

An important dimension of fandom is that it can offer individuals a sense of belonging, even if they do not interact with other members of that group (Sandvoss 2005). Aca-fans' relationship to other fans is complicated when studying not just the object text but the fandom itself. As Katherine Larsen and Lynn Zubernis (2011, 45) highlight, aca-fans are situated precariously between "us" and "them" within fan culture. Additionally, we are often dealing with our own "shame issues" about not just the consumption of popular culture but the excessive emotions or passions elicited by our own fandom, which are considered anathema to academic objectivity. Larsen and Zubernis (2011, 45) observe, "As a result we theorise and politicise our pleasures in order to make them more palatable to a cultural elite that does not need any more encouragement to dismiss what we study as frivolous and meaningless." Yet, in giving these shameful pleasures space within the academe and other critical conversations, aca-fans can assist others in understanding the important social, cultural, economic, and political significance of the texts that we consume every day—or for one week in May—alongside their affective pleasures without necessarily privileging one over the other. Bertha Chin and Lori Hitchcock Morimoto (2013; cf. Iwabuchi 2010) caution against perpetuating the false binary often made between scholarship of the socio-political implications of fandom, positioned a "good", and the affective meanings and pleasures of fandom, positioned as "bad".

Eurovision fandom problematises this false binary as socio-political meaning and affective pleasures are frequently intertwined. Not only can fans switch between a discussion of the pleasure they take from Eurovision to a critical analysis of its political meanings, the pleasure can be directly connected to the latter as many Eurovision fans take pleasure out of the contest's appeal to global affect. While the official rationale for Eurovision is the production of a television spectacle that also stimulates popular music production in Europe (O'Connor 2015, 6), its grand narrative is that it is the song contest that unites Europe (Yair 1995; Raykoff and Tobin 2007, 2–3; Tragaki 2013, 2–3; Fricker and Gluhovic 2013, 3). The contest's growing association with LGBTIQA representation and politics has also nurtured a broadly appealing narrative around values of diversity and acceptance. For many, Eurovision unsettles Alan McKee's (2016, 33) definition of fun as "pleasure without purpose"; rather, Eurovision is pleasure *as* purpose.

The paradox of Eurovision is that it is a competition between nation-states that offers a narrative of unity and collective identity, which has led to many comparisons to events like the Olympic Games (Baker 2017). The extent to which these ideas of unity and collective identity are enacted in any meaningful way on a national or supranational level is questionable, but Eurovision fans themselves frequently exhibit what Jenkins (2006, 154) terms "pop cosmopolitanism", referring to "the ways that the transcultural flows of popular culture inspire new forms of global consciousness and cultural competency." The song contest gives fans access to international popular culture, and from this they develop deeper interests in the cultures, languages, and politics of other places (Georgiou 2008). Furthermore, the identity of Eurovision fan, or "Eurofan" as it is self-designated in the twittersphere, offers a transcultural identity centred on the contest, its music, and its values.

Such characteristics are exhibited by Australian fans, but must be read through the lens of Australian multiculturalism, not in the least because it is this policy that led to the establishment of SBS as a national multicultural broadcaster that screens Eurovision. Additionally, Australian multiculturalism arguably equips Australians with some of the cultural competencies Jenkins refers to, priming them to receive these transcultural flows of popular culture. Many of these arrive relatively organically in Australia through the flow of people but are then nurtured through policy and media strategy—such is the case of Eurovision. Popular culture as multiculturalism has typically been denigrated as offering only a trivial or weak form of cosmopolitanism, but the advent of scholarship in everyday multiculturalism has helped redeem this. Everyday multiculturalism is a conceptual and methodological approach to understanding multiculturalism not as a top-down policy but as something constituted by the lived experiences and everyday practices of those living in culturally diverse societies (Harris 2009; Wise and Velayutham 2009). By acknowledging the importance of the mundane, including interactions with media, everyday multiculturalism legitimises popular culture as a space of intercultural encounter (Harris 2009, 194). Read between Jenkins' pop cosmopolitanism and everyday multiculturalism, Eurovision is a valuable platform for intercultural encounter enjoyed by Australians and Eurofans abroad.

To understand the emergence of Australia as a significant, non-European audience for the Eurovision Song Contest, it is important to start at the beginning of the contest's broadcast in Australia. Accordingly, Chapter 2 delves into the institutional history of the Special Broadcasting Services (SBS) to uncover the broadcaster's rationale for screening Eurovision as part of its multicultural programming remit in the 1980s, connecting this to the role that the song contest has played for the EBU and its Eurovision media service. In order to develop Eurovision into the cult subcultural event it is for hundreds of thousands (occasionally millions) of Australian viewers it is each year, SBS and its production partner BlinkTV have added localised elements of its production, such as commentary and secondary programming. To further illustrate this, Chapter 3 focuses upon the important interpolative role commentary plays for a national audience, while also assessing how Australian Eurovision fans assess the role and performance of their commentators. Importantly, commentary can assist fans in feeling a greater sense of belonging to the contest.

The majority of participants in this study, including interviewees, first began watching the song contest in the 2000s (37%), with the second highest recruitment, so to speak, of fans occurring in the 2010s (21.7%), and the third highest in the 1990s (17.7%). The remaining fans began watching in the 1980s (12.9%), with small numbers watching abroad in the 1970s (6.4%) and 1960s (2.4%), and a few being unable to recall precisely. Family and friends play a shifting role in why people first started watching Eurovision over the years, but many also report a sense of personal curiosity, often further motivated by word of mouth or media coverage. Family and serendipity (that is, stumbling across it in some manner, usually while channel surfing) were the prevalent reasons amongst those who started watching in the 1980s and the 1990s, while recommendations from friends and curiosity increase in prevalence throughout the 1990s, 2000s, and 2010s. This can be attributed to several factors, including the growth of SBS's reputation for 'cult' rather than 'family' programming in the 1990s and 2000s and its active promotion of Eurovision as a social event to be enjoyed with friends at a screening party. Participants who started watching in the 2010s make almost no mention of the role of family in being first motivated to watch Eurovision; friends or partners, social media, and sheer curiosity at the phenomenon are the most prevalent reasons; a very small minority nominate Australia's entrance into the competition as their motivation for watching. The fourth chapter engages with shifting ideas of family, belonging, and community in a changing national and global context for Eurovision, with a particular focus on the evolution of the Eurovision party as a particular type of communal celebration.

Having established the important media and social role played by Eurovision in Australia, Chapter 5 interrogates its function within multicultural Australia as a means of connecting to European heritage, examining also how Australians understand Eurovision as an intercultural and transnational encounter. The Eurocentrism of the dominant 'European connection' narrative about Eurovision in Australia is problematic as it idealises a culturally diverse but racially homogenous image of Australia that does not hold with its contemporary reality. While SBS is instrumental in perpetuating the European connection narrative, their casting and production of Australian Eurovision content disrupts this by foregrounding Asian and Indigenous faces in the self-conscious act of representing Australia on an international stage.

The sixth and seventh chapters broaden the analysis to incorporate the political and cultural impacts of Australia's participation in the Eurovision Song Contest since 2015. Chapter 6 provides an examination of how Australia has been represented as a nation on this international stage through an analysis of the Australian performances and international responses to Australia's participation. To conclude, Chapter 7 reflects upon Australia's role in the global expansion of the Eurovision concept. After the announcement that Australia would be participating in the contest for a second time, it was also revealed that SBS and the EBU were planning an Asia-Pacific version of the contest called Eurovision Asia, but referred to by fans colloquially as "Asiavision". Fan responses to the prospect are mixed, revealing not only ambivalent attitudes towards Australia's geopolitical status, but also articulating a strong sense of what Eurovision can and should be in this globalised, twenty-first century context.

References

Baker, Catherine. 2017. "The 'Gay Olympics'? The Eurovision Song Contest and the Politics of LGBT/European Belonging." *European Journal of International Relations* 23 (1): 97–121.

Bourdon, Jérôme. 2007. "Unhappy Engineers of the European Soul: The EBU and the Woes of Pan-European Television." *International Communication Gazette* 69 (3): 263–280.

Chin, Bertha, and Lori Hitchcock Morimoto. 2013. "Towards a Theory of Transcultural Fandom." *Participations: Journal of Audience and Reception Studies* 10 (1): 92–108.

Eberhardt, Maeve, and Kara Freeman. 2015. "'First Things First, I'm the Realest': Linguistic Appropriation, White Privilege, and the Hip-Hop Persona of Iggy Azalea." *Journal of Sociolinguistics* 19 (3): 303–327.

EBU. 2018. "EBU Terminates This Year's Partnership with Mango TV." *Eurovision.tv*, May 11. https://eurovision.tv/snippet/ebu-terminates-this-year-s-partnership-with-mango-tv.

Ferreira, Fernando, and Joel Waldfogal. 2013. "Pop Internationalism: Has Half a Century of World Music Trade Displaced Local Culture?" *The Economic Journal* 123: 634–664.

Fiske, John. 2002. "The Cultural Economy of Fandom." In *The Adoring Audience: Fan Culture and Popular Media*, edited by Lisa A. Lewis, 30–49. London: Routledge.

Fricker, Karen, and Milija Gluhovic. 2013. "Introduction: Eurovision and the 'New' Europe." In *Performing the 'New' Europe: Identities, Feelings and Politics in the Eurovision Song Contest*, edited by Karen Fricker and Milija Gluhovic, 1–28. Houndmills: Palgrave Macmillan.

Frith, Simon. 2011. "Pop Music." In *The Cambridge Companion to Pop and Rock*, edited by Simon Frith, Will Straw, and John Street, 91–108. Cambridge: Cambridge University Press.

Georgiou, Myria. 2008. "'In the End, Germany Will Always Resort to Hot Pants': Watching Europe Singing, Constructing the Stereotype." *Popular Communication* 6 (3): 141–154. https://doi.org/10.1080/15405700802198188.

Gripsrud, Jostein. 2007. "Television and the European Public Sphere." *European Journal of Communication* 22 (4): 479–492.

Harris, Anita. 2009. "Shifting the Boundaries of Cultural Spaces: Young People and Everyday Multiculturalism." *Social Identities* 15 (2): 187–205.

Hay, Chris, and Billy Kanafani. 2017. "Boos, Tears, Sweat, and Toil: Experiencing the 2015 Eurovision Song Contest Live." *Popular Entertainment Studies* 8 (1): 57–73.

Iwabuchi, Koichi. 2010. "Undoing Inter-national Fandom in the Age of Brand Nationalism." *Mechademia* 5 (1): 87–96.

Jenkins, Henry. 1992. *Textual Poachers: Television Fans & Participatory Culture, Studies in Culture and Communication*. New York: Routledge.

Jenkins, Henry. 2006. *Fans, Bloggers, and Gamers: Exploring Participatory Culture*. New York and London: New York University Press.

Jenkins, Henry. 2008. *Convergence Culture: Where Old and New Media Collide*. New York and London: New York University Press.

Knox, David. 2017. "Sunday 14 May 2017." *TvTonight*, May 17. http://www.tvtonight.com.au/2017/05/sunday-14-may-2017.html.

Larsen, Katherine, and Lynn Zubernis. 2011. *Fandom at the Crossroads: Celebration, Shame and Fan/Producer Relationships*. Newcastle upon Tyne: Cambridge Scholars Publishing.

McKee, Alan. 2016. *Fun! What Entertainment Tells us About Living a Good Life*. Houndmills Basingstoke Hampshire: Palgrave Macmillan.

O'Connor, John Kennedy. 2015. *The Eurovision Song Contest: The Official Celebration*. London: Carlton.

Phillips, A. A. 1950. "The Cultural Cringe." *Meanjin* 9 (4): 299–302.

Raykoff, Ivan, and Robert Deam Tobin. 2007. "Introduction." In *A Song for Europe: Popular Music and Politics in the Eurovision Song Contest*, edited by Ivan Raykoff and Robert Deam Tobin, xvii–xxi. Aldershot: Ashgate.

Reuters. 2018. "Turkey Boycotts Eurovision Song Contest over LGBTQ Performers." *Huffington Post*, August 9. https://www.huffingtonpost.com/entry/turkey-eurovision-boycott_us_5b6c641de4b0bdd062074ac2.

Sandvoss, Cornel. 2005. *Fans: The Mirror of Consumption*. Cambridge: Polity.

Singleton, Brian, Karen Fricker, and Elena Moreo. 2007. "Performing the Queer Network: Fans and Families at the Eurovision Song Contest." *SQS–Suomen Queer-tutkimuksen Seuran lehti* 2 (2): 12–24.

Times of Israel. 2018. "Turkish PM Claims Israel's Eurovision Win Is Part of an Imperialist Plot." *Times of Israel*, June 16. https://www.timesofisrael.com/turkish-pm-claims-israels-eurovision-win-part-of-an-imperialist-plot/.

Tragaki, Dafni. 2013. "Introduction." In *Empire of Song: Europe and Nation in the Eurovision Song Contest*, edited by Dafni Tragaki, 1–33.

Vuletic, Dean. 2018. *Postwar Europe and the Eurovision Song Contest*. London: Bloomsbury.

Wise, Amanda, and Selvaraj Velayutham. 2009. "Introduction: Multiculturalism and Everyday Life." In *Everyday Multiculturalism*, edited by Amanda Wise and Selvaraj Velayutham, 1–17. Houndsmills: Palgrave Macmillan.

Wooton, Dan. 2016. "Brit Eurovision Host Graham Norton Slams Aussie Entry and Calls for the Country to Be Banned from the Contest." *The Sun*, May 9. https://www.thesun.co.uk/archives/bizarre/1168044/brit-eurovision-host-graham-norton-slams-aussie-entry-and-calls-for-the-country-to-be-banned-from-the-contest/.

Yair, Gad. 1995. "'Unite Unite Europe': The Political and Cultural Structures of Europe as Reflected in the Eurovision Song Contest." *Social Networks* 17 (2): 147–161.

CHAPTER 2

A Tale of Two Broadcasters: The EBU and SBS

Abstract The Eurovision Song Contest first emerged in the 1950s with the pragmatic goal of promoting the European Broadcasting Union's new "Eurovision" media sharing service, but the loftier aim to create a share European identity and public sphere has since dominated most understandings of its history. In Australia, the song contest is screened on the Special Broadcasting Service (SBS), a hybrid public/commercial broadcaster charged with a remit to serve and represent multicultural Australia. Just as the EBU sought to create a European public sphere, SBS's role in the Australian television landscape was to create a multicultural Australian public sphere that was also globally minded. The song contest's popularity in Australia is both a contributing factor and resulting product of this.

Keywords Broadcast television · Cosmopolitan habitus · Multicultural policy · Public sphere · Media

To understand the longevity and appeal of the Eurovision Song Contest for its Australian audiences, it is important to understand how the contest first emerged and how it came to be on Australian screens in the first place. This history is inextricably tied to the rise of television—specifically public television—as an important national and international medium of entertainment and communication and its development into a global industry and culture into the twenty-first century.

J. Carniel, *Understanding the Eurovision Song Contest in Multicultural Australia*,
https://doi.org/10.1007/978-3-030-02315-7_2

It is also tied to the cultural and ideological reconstruction of Europe after World War II and in the midst of the Cold War. Such observations hold true even in the Australian context: the song contest is part of Australia's broadcast history, specifically the development of the Special Broadcasting Services (SBS), and the development of multicultural policy and ideology as part of its official reconfiguration as a culturally diverse nation in the wake of post-World War II mass migration, which until the 1970s was predominantly from Europe. The Eurovision network and its song contest was intended to foster a sense of cultural connection between diverse European nations. SBS was intended to assist migrants in feeling a connection to both their new and old homelands, as well as fostering a stronger sense of a culturally diverse Australian national image for other non-migrant viewers. Programming the Eurovision Song Contest in Australia contributed to SBS's mandate to assist migrants in maintaining a sense of connection to European homelands while also introducing non-migrant and non-European Australian viewers to a new cultural experience, potentially uniting them through this shared televisual experience. Eurovision thus contributed to two separate projects that sought to unify fragmented imagined communities bound by either national or regional identity through the use of media, communications, and culture.

While television technology has a longer history and broadcasters first start emerging in the 1920 and 1930s, television began to be a popular cultural medium after World War II, particularly in the 1950s when the effects of economic reconstruction made it more affordable and accessible, at least for the middle classes. Concurrent with this in the European context is the development of transnational media services with the pragmatic aim of sharing media resources and the loftier aim of fostering a shared European culture and identity that would support other regional initiatives, such as the European Economic Community and the Council of Europe. The European Broadcasting Union (EBU) was developed in 1950 as an agreement between twenty-three public broadcasters, and was preceded by the International Broadcasting Union (IBU). While the broadcasters involved were public, the EBU differed from the IBU and other preceding broadcasting unions in that it was formed by an agreement between the broadcasters themselves rather than states that funded them, and that it centred upon mediating technical conflicts as well as facilitating an exchange of programming. Jostein Gripsrud (2007, 285) emphasises the importance of the EBU in developing a kind of

European public sphere, in the Habermasian sense, through the medium of television. To do so, it developed programs specifically designed to give Europeans shared televisual experiences that might also foster a sense of belonging to an imagined community, to extrapolate Benedict Anderson's (2006) idea into a regional context. The EBU's Eurovision Song Contest was thus conceived with two intentions: to unite the nations of post-war Europe through a shared cultural event, and to promote the EBU's Eurovision distribution network (O'Connor 2010, 8).

The label "Eurovision" is now synonymous with the song contest rather than the EBU's broadcasting initiative, but the original intention was that the programming—and the song contest as but one part of this—would unite Europe through a shared televisual culture that could "forge a new collective conscience and help the new Europe supersede old nations" (Bourdon 2007, 265). The song contest was part of a broader schedule of programming that included dramatic adaptations of European literary classics and plays, sports, and news services. The success of this programming was varied. Jerome Bourdon suggests that the most successful ventures were those that were seen and celebrated by the audiences as "Euro-events" (Gripsrud 2007, 485); that is, specific live telecasts that could be experienced as pan-European events, such as the Eurovision Song Contest, rather than programs, such as the mini-series and plays, that were still too firmly located within particular national-cultural canons. Indeed, live television proved to be an integral element to any success for the EBU's project as it assisted in the creation of a sense of community that was bound by a shared experience across several time zones and other usual organising principles. In *Eurovision! A History of Modern Europe Through the World's Greatest Song Contest*, Chris West (2017) suggests that musical taste also played an important role in helping to define Eurovision's role in articulating a European identity. Eurovision targeted a family audience shaped by more traditional tastes, and for decades remained largely untouched by diverse and new music styles that were popular with modern youth, such as the nascent genre of rock'n'roll.

The song contest itself was the brainchild of Marcel Bezençon, director of the EBU for its first twenty years. He took his inspiration from the *Festa della canzone italiana di Sanremo* (the Sanremo Music Festival), which begin in Italy in 1951 and is still ongoing. Eurovision began in Lugano, Switzerland, in 1956 with a modest seven participating nations, each presenting two entries. The UK, represented in the EBU by the

BBC, did not participate in the inaugural contest, committing instead to its own program, *Festival of British Popular Songs*, which ran for only two years in 1956–1957. This contradicts a popular fan myth, since debunked by Eurovision staff writers Gordon Roxburgh and Paul Jordan (2017), that the BBC was absent only because they missed the participation deadline. Yet West (2017, 15) insists that Britain did have an entrant at the ready—Australian ex-pat performer Shirley Abicair—but is not clear why her entry was never finalised. In contrast to the extended public voting spectacle in the contemporary contest, votes were cast privately, so it is unknown who voted for whom, but West (2017, 14) notes that there were more Swiss judges on the panel than from the other nations. Switzerland's Lys Assia won with her second song, "Refrain" and the only video footage remaining of the first contest is her winner's reprise; fortunately, the audio recording of the radio broadcast is available.

Eurovision has since expanded at a fairly steady rate, from seven participating nations in 1956 to forty-three in 2018. Participants are drawn from EBU membership, which now boasts 73 broadcasting companies from 56 nations as full members, and 33 broadcasters from 21 countries as associate members. Full membership is determined largely by the logistics of geography but, as Cornel Sandvoss (2008, 205) suggests and as the acceptance of the Australian bid to participate attests, it is also driven by purported shared European (or Western) values. Australia has been represented in the EBU associate membership since 1950 by the Australian Broadcasting Corporation (ABC), which had distribution rights for the radio broadcast of the original contest in 1956 (Vuletic 2018, 49). While the ABC's connection to Eurovision has been revitalised recently as its children's channel, ABC Kids, is now home to the Junior Eurovision Song Contest, SBS is the Australian broadcaster most readily associated with EBU content, such as the song contest, global news, and international sports coverage (particularly the FIFA World Cup). SBS joined the EBU in 1979, when it was in the midst of its transformation from a community radio service into a quasi-national television and radio broadcaster.[1]

[1]It is deemed "quasi-national" at the historical juncture described here as SBS was slowly rolled out across the Australian continent, starting with Sydney and Melbourne in 1980, and arriving in Adelaide, Brisbane, Perth and Hobart in 1986. Darwin was the last capital city to receive SBS in 1994. It is now available nationally.

Although it is now a staple in the Australian media landscape, SBS is a relatively young broadcaster in comparison to the EBU and the ABC. Where the ABC has a broad remit to represent and foster Australian national culture, SBS operates under a charter in which its principle function is "to provide multilingual and multicultural radio and television services that inform, educate and entertain all Australians and, in doing so, reflect Australia's multicultural society" (SBS, n.d.). It was established in 1980 as part of the Australian federal government's commitment to developing significant multicultural infrastructure to serve the needs of its increasingly diverse population but, as Ien Ang, Gay Hawkins and Lamia Dabboussy's (2008, 272) history of SBS has revealed, the network has evolved into a media service that has been most innovative in meeting the challenges of cultural diversity, "global and local, national and transnational." Today, SBS not only represents local iterations of Australian multiculturalism, but also actively promotes transnational and globalised connections between Australia and the rest of the world.

SBS has never been particularly conservative or homophobic in its programming but in recent years it has made a concerted effort to incorporate sexuality as part of its remit to represent Australian diversity (Ang et al. 2008, 142) by increasing queer programming, including the telecast of the Sydney Gay and Lesbian Mardi Gras since 2014, and establishing an area on its website to discuss sexuality issues and news. Furthermore, its former Managing Director (2011–2018), Michael Ebeid, is openly gay, as are several of its news presenters. These public figures facilitate a strong representation for multicultural queerness in the national space, particularly in comparison to other broadcasters.

Unlike the ABC, SBS is a hybrid of public and commercial television, as it has necessarily subsidised its public funding with advertising since 1991, increasing the frequency and duration of its ad breaks in 2006. SBS has been criticised for becoming more commercially orientated in this process; together with its increasingly global-oriented programming, exemplified by its 2016 deal with North American VICE media, this is seen as a significant and problematic move away from its multicultural remit (Mullins 2010; Enker 2004). Yet Belinda Smaill (2002, 396) notes that almost from its outset, SBS's programming required a fairly pragmatic and mainstream approach in order to ensure its appeal to its entire national audience, and that even when SBS's programming was seen as better serving its minority communities, it was subject to criticism for both ghettoisation *and* a privileging of minority interests over those of "mainstream Australia" (Smaill 2002, 397; see also Flew 2009). Evidently, SBS cannot escape criticism, regardless of the direction it takes. Although not all its strategies are defensible, its global shift, which does come also with increased commercialisation, allows for a reflection of Australia's diversity while maximising transnational connections.

The Eurovision Song Contest has broadcast on SBS since 1983. Although the song contest is absent from Ang et al's history of the broadcaster, it nevertheless constitutes an important transnational cultural connection offered by SBS. It neatly negotiates some of the tensions of multicultural and mainstream success, and through this has been pivotal in acculturating Australians into a multicultural norm, which Gay Hawkins and Ien Ang (2007) describe as a "cosmopolitan habitus". They argue that while multicultural policy provides important "big picture background" for the emergence of SBS and its programming, the network was also responding to possibilities in Australian television that were being ignored by the ABC and the commercial networks. Using case studies of subtitling and world news, Hawkins and Ang (2007, 10) demonstrate how SBS used these innovations and its broader programming to "exploit and manage foreignness", as well as cultural anxieties about foreignness. In so doing, it created new publics and new uses for television in Australia, and facilitated Australian participation an "international public sphere." To some extent, this is an expansion of the same conceit on which the Eurovision network itself functioned—or at least that it sought to encourage, as suggested by Gripsrud's notion of a European public sphere; that is, it presumes a European viewer interested in regional events, be they political or cultural, as a means of fostering a unified regional identity. Similarly, screening Eurovision in Australia presumes the viewer's global, or at least European, interest. Viewed in this light, SBS is not limited to serving a migrant audience, but inhabits a far more expanded and significant role in "combatting cultural insularity and encouraging a more cosmopolitan habitus" (Hawkins and Ang 2007, 6) through making cultural otherness both accessible and normalised.

This interpretation of the SBS project has numerous implications for understanding the place of Eurovision on the network and for Australian audiences. Specifically, it asks us to consider how the Eurovision programming aligned with SBS's objectives to provide a new form of television that was also a translation of specific policy objectives. Certainly, there was nothing like Eurovision on Australian television, a handful of talent shows notwithstanding; indeed, many would agree that even with the proliferation of music reality/talent shows, there remains nothing else like it today. At the time that Eurovision first started screening in Australia, the contest was under the so-called language rule, wherein the EBU had decreed that all songs were to be sung in one of the official languages of the competing country. Eurovision at this time fulfilled the SBS remit to provide a multilingual service, and it was perhaps the only truly multilingual program that did not feature subtitles.[2] While Hawkins and Ang suggest that it is through the presence of subtitles that SBS fostered a cosmopolitan habitus wherein viewers were encouraged to embrace global and linguistic diversity, it was the *absence* of subtitles in the Eurovision broadcast that facilitated this.

A romantic argument might be made here for music as a unifying language beyond its lyrics, and certainly this is part of the ongoing mythology surrounding the contest, but it is important to note that the contest itself was hosted in English and French as the official languages used by the EBU. This enabled Anglophone (and Francophone) viewers in Australia to understand what was happening, regardless of their first language; in the case of the 1991 contest, which flouted the requirements regarding use of French and English by being hosted almost entirely in Italian, Sir Terry Wogan's English-language commentary provided de facto interpretation. Furthermore, all viewers, migrant and non-migrant alike, heard a variety of languages in the songs themselves, some which the viewers might connect with personally and culturally, but many of which were likely unfamiliar. Even with individual moments of linguistic comprehension and connection, all viewers were engaged in what Hawkins and Ang characterise as a democratic linguistic viewership because of shared moments where all engaged with unfamiliarity and foreignness. Yet it is important to remember that the specific transmission for much of this time was British, which comes with its own complex relationship to ideas of foreignness; while Wogan's commentary was appreciated for its wit, he was considered increasingly xenophobic in the later years.

This is, however, not a particularly Australian experience as these shared moments of familiarity and foreignness occur in both the European and Australian contexts. The notion of an Australian Eurovision audience is established through their shared time and place, which for much of the history of the contest was drastically different to the European temporal and emplaced experience of the television event.

[2] In recent years, subtitles have been added to the delayed broadcast.

Watching a delayed broadcast on a Sunday evening on a national broadcaster that was at once Australian and yet internationalist or cosmopolitan in scope became particular to the Australian experience of Eurovision. Even though Australians can now share the experience contemporaneously with European viewers, the early start on a late autumn morning still marks the Australian experience as different from the European.

In reviewing the origins and history of the Eurovision Song Contest as a means of promoting a new international media network, it is tempting to take a cynical perspective about the exploitation of desirable values by corporate entities (acknowledging, of course, that the EBU is a not-for-profit organisation). Yet what fan studies as whole demonstrates through the concept of participatory culture is the way that texts become something else in the hands of those that consume them (Jenkins 1992). Continuing cynically, perhaps fans have bought into the EBU narrative of a song contest that unifies, but in 1956 even those media visionaries involved in the development of Eurovision could not have predicted the development of a global network of fans who have taken on the purported values of Eurovision and, through their own activities and interpretations, have made them a reality. Eurovision fans believe in the song contest as a way to celebrate unity, diversity, and music, and so this is what it has become. Certainly, what Marcel Bezençon and his colleagues could not have predicted is that the contest would have grown to include Australians not just as another audience, but as active participants.

As the song contest has evolved into the twenty-first century, it has become an important part of cultural globalisation. While it is a shared televisual experience, it is interpolated locally through localised media, branding, and productions. Even prior to Australia's entry into the contest itself, the Eurovision broadcast has emerged as a core element of the SBS brand in the twenty-first century. With the retirement of Wogan in 2008, SBS were able to take advantage of this changing of the guard to implement local commentary better suited to the tone and ethos of SBS and its viewers. This is the culmination of a concerted effort throughout the 2000s to develop and better cater to the self-consciously Australian audience that began to emerge and is a crucial element to how Eurovision rose to greater prominence in the Australian television viewing calendar. It is now a complex text that caters to the tastes of a diverse, cosmopolitan audience who value a form of entertainment that is inclusive but still different to what is on offer on mainstream television channels in Australia.

References

Anderson, Benedict. 2006. *Imagined Communities: Reflections on the Origin and Spread of Nationalism*. Rev. ed. London: Verso.

Ang, Ien, Gay Hawkins, and Lamia Dabboussy. 2008. *The SBS Story: The Challenge of Diversity*. Sydney: UNSW Press.

Bourdon, Jérôme. 2007. "Unhappy Engineers of the European Soul: The EBU and the Woes of Pan-European Television." *International Communication Gazette* 69 (3): 263–280.

Enker, Debi. 2004. "Where to Now, SBS?" *Age*. Accessed November 21, 2016. http://www.theage.com.au/articles/2004/05/26/1085461820488.html.

Flew, Terry. 2009. "The Special Broadcasting Service After 30 years: Public Service Media and New Ways of Thinking About Media and Citizenship." *Media International Australia* 133 (1): 9–14.

Gripsrud, Jostein. 2007. "Television and the European Public Sphere." *European Journal of Communication* 22 (4): 479–492.

Hawkins, Gay, and Ien Ang. 2007. "Inventing SBS: Televising the Foreign." *Australian Cultural History* 26: 1–14.

Jenkins, Henry. 1992. *Textual Poachers: Television Fans & Participatory Culture, Studies in Culture and Communication*. New York: Routledge.

Mullins, Michael. 2010. "Multiculturalism Steps Aside for Advertising on SBS." *Eureka Street* 20 (12): 52–53.

O'Connor, John Kennedy 2010. *The Eurovision Song Contest: The Official History*. London: Carlton.

Roxburgh, Gordon, and Paul Jordan. 2017. "Shining a Light on the United Kingdom: 60 Years at Eurovision." *Eurovision.tv*, January 12. https://eurovision.tv/story/shining-a-light-on-the-united-kingdom-60-years-at-eurovision.

Sandvoss, Cornel. 2008. "On the Couch with Europe: The Eurovision Song Contest, the European Broadcast Union and Belonging on the Old Continent." *Popular Communication* 6 (3): 190–207.

Special Broadcasting Service (SBS). n.d. *SBS Charter*. Accessed July 26, 2015. http://www.sbs.com.au/aboutus/corporate/index/id/25/h/.

Smaill, Belinda. 2002. "Narrating Community: Multiculturalism and Australia's SBS Television." *Journal of Communication Inquiry* 26 (4): 391–407.

Vuletic, Dean. 2018. *Postwar Europe and the Eurovision Song Contest*. London: Bloomsbury.

West, Chris. 2017. *Eurovision: A History of Modern Europe Through the World's Greatest Song Contest*. London: Melville House.

CHAPTER 3

Aussie-Fying Eurovision: Local Commentary as Media Interpolation

Abstract Local commentary of the contest plays an important role in interpolating the Eurovision Song Contest for specific national audiences. For its first twenty-five years, Australian audiences received the BBC broadcast with commentary by the late Sir Terry Wogan, introducing consistent Australian commentary from Sam Pang and Julia Zemiro upon Wogan's retirement in 2008. Fans value knowledge, passion, and humour from their commentators, which have been delivered by Australian commentators with various degrees of success. This chapter explores these mixed views of Australian Eurovision commentary with a particular consideration of how multiculturalism and queerness are utilised within this national context.

Keywords Commentary · Media interpolation · Multiculturalism · Queerness · Humour

The vast majority of the participants in this study first began watching Eurovision in the early 2000s and it is easy to see how the contest increased in appeal throughout this time, particularly for younger generations. External to the Australian context, the contest itself had commenced a populist evolution in the late 1990s. The live orchestra was dismissed, the live, fan-driven audience became an important production element, and contemporary music trends, specifically dance music, began to feature

J. Carniel, *Understanding the Eurovision Song Contest in Multicultural Australia*,
https://doi.org/10.1007/978-3-030-02315-7_3

more prominently in the offerings. Coinciding with this, throughout the 1990s SBS began framing its content as cosmopolitan and cultish rather than minoritarian; with its appeal to SBS's twin audiences of "ethnics" and "eggheads" (Ang et al. 2008) Eurovision effectively bridged this divide. In Australia, the contest was increasingly marketed towards a youthful, cosmopolitan, and technologically savvy audience. SBS began to actively foster the idea of an Australian Eurovision subculture throughout the 2000s, achieving this through the implementation of online polling (in lieu of an official vote in years prior to 2015), the provision of Eurovision party resources via its various media outlets, such as *Feast* magazine and the SBS website, and its development of a significant social media presence on Facebook and Twitter. Most importantly, SBS began to experiment with locally supplied commentary and other domestic productions, such as preview shows, documentaries, and other specials. In so doing, SBS transitioned itself from merely a means of relaying the event to an active producer of content. This is exemplified by their introduction of local Australian commentary. The further effect of these Australian production interventions is the thorough integration of Eurovision into Australian television culture, and of Australians into Eurovision culture.

The introduction of Australian commentary is popularly remembered by fans as the entry of Sam Pang and Julia Zemiro in 2009 rather than Effie (Mary Coustas) or Des Mangan who each made an attempt in the early 2000s, perhaps because of Pang and Zemiro's relative success and longevity. These various interventions into the commentary throughout the 2000s are noted by fans as when "Australia" started to take Eurovision more seriously because SBS were seen to be taking it more seriously by actively investing in its production. Commentary was an important way for SBS to mark the product as, in part, their own, therefore also something that belongs to Australia. As one fan observed, "[local commentary] makes our country really feel part of it all - we're not just sad Aussies with a strange hobby, a national TV broadcaster sends its own people there to cover it!" Another agrees that "there's something special about listening to your own country's comments."

The integration of Australian commentary was not as smooth as fan memories might suggest, as there were two other attempts to implement this before Zemiro and Pang were able to settle into their roles with great success. SBS initially used Sir Terry Wogan's famed commentary for the BBC, but in 2001 it experimented with local commentary from Mary Coustas in the character of Effie, an exaggeration of Greek

Australian femininity who first found popularity via the television show, *Acropolis Now!* on the commercial Seven network. After the failure of Effie's commentary, SBS returned to Wogan and the BBC for another year before attempting their own coverage again, this time with cult movies host Des Mangan. For better or worse, Wogan had become part of Australia's Eurovision tradition. When Wogan stepped down from BBC coverage at the end of 2008, fans had to accept that a new commentator was a practical reality. This was an opportunity for SBS to create a new Australian Eurovision tradition in a far more seamless manner than previous attempts. In addition to their local popularity, Zemiro and Pang established themselves as a known Australian presence in the Eurovision media rooms and backstage. They were replaced in 2017 by Myf Warhurst and Joel Creasey, who were initially received with some ambivalence by fans, largely because of the immense love and familiarity many expressed for their predecessors.

Event commentary is something most readily associated with sports broadcasting. Specific similarities to Eurovision can be drawn in terms of characteristics of competition and national representation that are central to both the song contest and international sporting events; indeed, numerous scholars suggest that Eurovision fandom share greater similarity with sports fandom than it does with any other television fandoms, such as those based around narrative series (Fricker et al. 2007; Georgiou 2008; Sandvoss 2008). In media scholarship, sports commentary is largely understood through the concept of framing, wherein material is presented in particular ways for interpretation by the viewer or consumer (Parker and Fink 2008). Framing determines and highlights what is important, and de-emphasises that which is deemed less important; yet omissions from a broadcast are equally telling. In televised events, framing is visually supported by editing, close-ups, and replays, while audio commentary provides further contextual cues to assist or direct the viewer in their interpretation of events. While Eurovision today involves a large live audience of between 7000 and 10,000 fans, the song contest is by genesis and definition a television event. As the broadcast is produced and edited by the EBU, the visual framing is determined by an organisation that claims to be apolitical and whose primary objective is to create a polished, successful, and entertaining television event that appeals to an international audience. Commentary therefore becomes important at the level of the national broadcast as this is where framing can be tailored to and directed at a specific national audience.

Commentary works to mediate the viewer's understanding of what is occurring on the stage and screen of the Eurovision production. Being localised—that is, provided by the relevant national broadcaster—it mediates the international spectacle through the national lens. Furthermore, being in the local language it may assist in literally translating the event and its songs. Commentary as a form of mediation of foreignness can be understood through the process of interpolation, wherein a media text is framed through the assumed subject position for its audiences. Hawkins and Ang's observations about the differences between SBS and ABC international news coverage is useful for understanding this process within the Australian context. With reference to the ABC's establishment of foreign news bureaus, the authors observe, "The ABC locates the significance of its international content in relation to viewer's assumed identification with Australia. It interpolates its viewers as Australians looking at the world. In this model, the national not only mediates the international but gives meaning to it" (Hawkins and Ang 2007, 9). Australian Eurovision commentary, like Australian international news bureaus, similarly interpolates viewers as Australians looking at the world. As an SBS production element, it also assumes a distinctly multicultural and cosmopolitan viewer who is comfortable with foreignness and diversity but still requires additional information for understanding the event and its participants. In addressing the commentary, fans indicate that commentators provide a particularly *Australian* way of looking at things, and frame the commentary as acts of cultural translation and contextualisation for the Australian audience. Furthermore, it enhances viewers sense of connection to the event.

The interpolation of Eurovision for Australian audiences has evolved as Australia's own relationship to the contest has shifted from observer to participant. Former commentator Julia Zemiro (quoted in Griffiths 2018) admits that this made her role "weird" as she felt it limited the way that she could relate to the competition: "all of a sudden you're in it, it's like going to someone's house and going to a dinner party and saying 'I don't really like the food'. You're not as free to be able to have fun with what you're watching because you're in it." A survey participant makes a similar observation: "For me it also removes the distinction between 'outsider/observer' and 'participant', which means that it's hard to watch the Contest in an impartial manner." Fans also observe a shift toward the "parochial" and "patriotic" in the broadcast approach; for example, one criticises Myf Warhurst for being *too* supportive of

Australian entrant Isaiah Firebrace. Jingoism within the Australian fan community is the focus of similar criticism: "There's too much blind following of the Australian acts regardless of quality and Australian fans are expecting good scores even though we are sending average songs." The shift in fan attitudes impacts expectations placed upon commentators; in contrast to the Warhurst example above, my own written commentary for the *Conversation* website was criticised by one reader for *not* being supportive enough of Firebrace (Carniel 2017). No longer simply the connected cosmopolitan, Australian viewers are now subsumed into the same nationalist competitiveness as other nations, thereby shifting the meaning of that assumed identification with Australia and the way this is expressed in commentary around the event.

Despite these issues, fans value local commentary because of the interpolative process but identify three elements that must be present in all forms of commentary regardless of origin and audience: knowledge, passion, and humour. The sequence here is important. While fans appreciate the use of humour in commentary, this is of tertiary importance to the provision of information, which in turn must be presented with accuracy, and a perceptible commitment to (or passion for) the subject material: "Commentary is best when the commentators don't talk too much, are humorous but not mean, sound like they are genuinely enjoying themselves and love being there, and most importantly, are actually informative, giving us information about the artists in the postcards." Subsequently, the greatest criticisms of commentators are lack of knowledge (or inaccuracy), lack of passion, and a disrespectful sense of humour. Those who consider themselves to be more serious fans are more likely to be ambivalent about commentary as a whole as their own knowledge and understanding may exceed that of the presenters; these fans will often watch the direct feed from the Eurovision website, which does not feature commentary. These serious fans acknowledge the usefulness of commentary for newer or more casual viewers but are wary of how the tone affects viewers understanding and appreciation of the contest. Wogan is identified by several fans as having an adverse effect on audience perceptions of the contest. "I feel he did a lot of damage to the reputation of the contest in the United Kingdom and elsewhere with his attitude to the show," said one participant. Another observes, "Terry Wogan is basically the reason that many people are "too cool" for Eurovision." A similar criticism is levelled against Wogan's successor on the BBC, Graham Norton, further emphasising the important duty commentators have in guiding the new or casual viewer: "Criticises everything and everyone. Just makes a huge joke of everything. I find it really rude. Unfortunately I feel anyone who is new to the show or doesn't know its history has the same opinion".

The fact that Wogan's commentary for the BBC was used by SBS for over twenty years is important to understanding the complexity of Australian attitudes to the contest, and the history of national interpolation for this audience. Wogan's commentary was, and remains in the memories of fans, a divisive characteristic of the BBC and SBS broadcasts. At best, Wogan's humour was dry and sardonic; at worst, he espoused views that were paranoid and xenophobic, emerging from a type of British exceptionalism from the continental European project perhaps best exemplified by the "Brexit" from the European Union. Nevertheless, Wogan (15%) emerges as a clear second to fan favourite Julia Zemiro (33%), and he is nominated by some as "the classic and the best." Australian fan criticisms of Wogan are consistent with those found elsewhere, such as Karen Fricker's (2013) close analysis of his commentary. While he embodied classic Eurovision, his later commentary was too subsumed in jingoism, xenophobia, and, in the words of one participant, "booze and bitterness". Fricker (2013, 54) argues that Wogan's commentary is "part of the phenomenon of populist Euroscepticism" that had been growing since the 1970s. Although Wogan increasingly deplored the apparent politicisation of Eurovision, his commentary also articulates anxieties about the shifting status of the UK in the changing regional and global landscape; the Eurovision Song Contest only served to underscore the UK's loss of status in this context. Central to Wogan's commentary was the positioning of the UK as external to Europe in any meaningful sense. As Fricker observes, Wogan constructed Eurovision as a "benignly strange, foreign ritual" that the UK participated in with some bemusement. Through his commentary "he efficiently interpolated a community of listeners who did not see themselves as part of the rest of Europe, nor as sharing its musical traditions and preferences" (Fricker 2013, 62). Furthermore, his assumed ideal viewer was one that shared Wogan's own identity profile as "male (or masculine-identified), white, and British (or British-identified), rather than European" (Fricker 2013, 64). This assumed viewer profile was already problematic in the British context, but is arguably more so once Wogan's commentary is exported to an Australian audience.

By using the BBC commentary, Eurovision was mediated by a British lens for an Australian audience, which further underscored the Anglocentric nature of Australian television culture and society that SBS was arguably intended to dilute or at least disrupt. Colonial Britishness formed part of Australian national identity, albeit problematically; indeed, it was not until 1984 that Australian citizens ceased to also be British subjects. Demographically, British descent was dominant in Australia for much of the twentieth century but the mass immigration programs of the post-World War II era changed Australia's cultural make-up significantly. In the decades immediately following the war, large numbers of migrants from southern and eastern Europe entered Australia through various immigration programs and it is in response to this changed demography and its impact upon the Australian media landscape that SBS itself emerges, and that Eurovision is introduced to Australian audiences more broadly (but, as will be explored in Chapter 5, this Eurocentric view of migration and Eurovision is problematic). While British Australians were also part of SBS's audience in Australia, their needs as migrants were not necessarily central to SBS's multicultural and multilingual remit, possibly because they constitute an "invisible migrant" group that blended into the racial and linguistic norm. Yet as Sara Wills (2004) has illustrated, blindness to actual cultural differences between Britain and Australia often led policy-makers and historians to overlook the fact that British migrants were still experiencing the same feelings of displacement and loss felt by other non-Anglophone migrant groups. While Australianness held a closer relationship to Britishness than other cultural presences due to the implantation of its institutions and language on to the Australian continent and nation, the two identities and cultures are not interchangeable.

Nevertheless, British perspectives were somewhat normalised on Australian television, perhaps more so than American texts. While these British texts are never considered Australian per se, in the 1980s they were more familiar to Australia's Anglophone audiences than those from, for example, France or Sweden. British humour in particular was prevalent on Australian television throughout the 1970s, 1980s, and 1990s, and remains popular to this day. Sue Turnbull (2010) links this to the presence of British radio comedies broadcast on ABC radio in the 1950s and the dominance of British sitcoms throughout Australian television history. British humour influenced the development of Australian humour (McKie and Natt 1996), and the two societies continue to exchange most cultural product with relative ease, but exchange is not tantamount to equivalency or sameness. Each has their own distinct cultural register and specificities, which manifests in the style of humour utilised in Eurovision commentary. Furthermore, Australian humour is also influenced by non-Anglo Celtic cultures, both migrant and Indigenous. John McCallum (2004, 204) suggests that Australia's history of migration, displacement, and dispossession creates a "very dry, laconic, self-mocking, cynical and irreverent" form of humour that is at once "a mixture of defiance and apology for being there," which perhaps best encapsulates Australian attitudes to its presence at Eurovision. Pang and Zemiro in particular were praised for bringing this distinctly Australian form of humour to their commentary in comparison to the British sensibilities of Wogan or his successor Norton, and were also seen to be appropriate representatives for a culturally diverse Australia and its multicultural broadcaster.

The success of Pang and Zemiro as a commentating team rests largely upon their particular chemistry, but it was also a matter of good timing. The departure of Wogan provided SBS with a valuable opportunity to make the most of fans' expectations for a changing of the guard. Previous attempts to introduce commentary were interruptions to the regular scheduled programming rather than an effective re-branding of the event for a distinctly Australian audience. Few participants in this study remember Des Mangan, but those who do praised him for his humour and knowledge of the competition—he also wrote a history of the contest, *This Is Sweden Calling* (2004). Even fewer fans remember Effie; indeed, even Mangan overlooks her in his book, expounding upon the importance of Australian commentary only in his account of his own entry into the role in 2003.

That Effie is considered by some fans as perhaps best forgotten is evidenced in her tying with Wogan for third least favourite commentator (after Joel Creasey and Graham Norton). Effie, played by Greek Australian actress Mary Coustas, was a popular character in the sitcom *Acropolis Now!* (1989–1992), based upon the *Wogs Out of Work* stage comedy show and also starring Nick Giannopoulos, George Kapiniaris, and Simon Palomares. The sitcom screened on the commercial Seven Network and was remarkable as one of the few (self) representations of multicultural Australia on mainstream television at the time, but it was also subject to criticism for its often uncritical use of stereotype: on the one hand the "wogs" characters were an act of comedic self-representation by second generation migrants (McCallum 2004, 212–214; Andreoni 2003, 90–91), but on the other hand they also fed into "anglo [sic], white, racist constructs of ethnicity" (Katsabanis and Murdolo 1993, 71). Coustas's Effie was one of the most popular characters to emerge from the show—she won a Logie for most popular comedy performer in 1993—and one of the most enduring, but she is also one of the most problematic. Maria Katsabanis and Adele Murdolo attempt to unpack their complex responses to the character of Effie as women of Mediterranean, non-English speaking backgrounds in their dialogic essay, "The World According to Effie." In so doing, they uncover the double bind of Effie, whereby laughing at her is racist but not laughing at her "privilege[es] some sort of essential Greek womanhood" (Katsabanis and Murdolo 1993, 73). Effie's popularity had waned somewhat by the time she became the Eurovision commentator, but she was still active on Australian television; that year, she also starred in a documentary series on SBS, *Effie: Just Quietly*, in which she used the brash naiveté of the character to investigate a range of social attitudes and practices.

The use of an ethnic stereotype as local commentator was a particularly contentious choice for the multicultural broadcaster. By casting Effie as the Eurovision commentator, SBS were attempting to play to a presumed audience of second generation migrant viewers through the use of a familiar second generation migrant character, but by doing so they were in effect placing Eurovision into the same symbolic field: an Othered stereotype of Europeanness worthy of satire. Yet, unlike the control wog humourists like Mary Coustas as Effie sought to exert over their representation, Eurovision was a powerless object of humour. While fans appreciate humour in their commentary, this comes with a strict caveat that commentators respect their subject matter. Two respondents frame their criticisms of Effie around this perceived lack of respect. One describes her commentary as making "a mockery of the event", while another connects her to the problematic place of kitsch in the contest: "It was a pisstake of what Eurovision [is]—I take the song contest seriously that's why I hated it when Lordi won for Finland in 2006—I'm a Eurovision purist I don't like the kitsch which has ebbed away since Loreen won for Sweden in 2012 [sic]." Independent of Eurovision, there is the possibility for complexity in the Effie character, as Katsabanis and Murdolo reveal, but within the context of the song contest this fails because it does not allow for it to be anything more than a laughable ethnic quirk.

There are other significant criticisms to be made of the 2001 production beyond the character of Effie. It was heavily edited to remove most of the content around the songs, such as postcards, hosts, and interval acts. Instead, SBS hosted a "studio party" (Douglas 2001) that combined footage from the contest with segments and commentary shot in their studios. Prominent multicultural Australians introduced songs from countries of their own background, queer cabaret performer Paul Capsis performed several Eurovision classics, and a panel of Australians discussed the entries. In an attempt at audience interactivity, viewers were encouraged to phone or email in with their nominations for best song, best/worst costume, best/worst dance routine, and so forth. Writing for the *Sydney Morning Herald*, cultural critic Adrian Martin (2001, 70) criticised the program's "super-camp, hyper-ironic Aussie sensibility" for alienating the contest's viewers and SBS viewers more broadly. Martin described the studio party as a "strange shotgun marriage between camp knowingness and a cultural conservatism", as the humour relied upon offensive cultural stereotypes, and not just in the form of Effie. Pressured by a deluge of complaints, SBS replayed the uncut broadcast two weeks later. Martin admonished, "I hope [SBS] has learnt a lesson about the perils of stereotyping and homogenising its highly diverse audience. Any attempt to pigeonhole people that brazenly is doomed to failure." The inclusion of Paul Capsis, who had at this time also garnered praise for his role as transgender, Greek Australian prostitute Toula in the 1998 film *Head On*, acknowledged the contest's growing queer audience but SBS's primary assumption about their audience was that they were "a second generation Australian who likes to mock Eurovision" (Douglas 2001). The incident reveals the complexity of identity politics and Eurovision fandom at the time. One SBS representative speaking on ABC radio characterised their critics as "Anglo-homophobes" (quoted in Douglas 2001), but the fans themselves vehemently resisted stereotyping and asserted their diversity. This diversity included those who were neither queer nor second generation Australians, who may have been "Anglo-homophobes", but who may have also belonged to the generic cosmopolitanised audience SBS had sought to cultivate throughout the 1990s. In contrast to the European connection narrative of today, wherein transnationalism is positioned positively, the framing of Effie's Eurovision studio party encouraged viewers to see the Eurovision Song Contest—and by extension the European culture and heritage it represented—as a relic of the Old World worthy of mockery.

This secondary SBS audience, characterised by Ang et al. (2008) as the "eggheads", appreciated the arthouse and cult offerings, such as foreign films, documentaries, and short films. Eurovision had emerged as one of these cult products but in their early interventions SBS made the error of focussing on the kitsch appeal rather than understanding its cosmopolitanising function in connecting Australians to popular music cultures in Europe. Not incidentally, their second local commentator Des Mangan was also the host of the *Cult Movie* show. In addition to being a self-confessed Eurovision fan, Mangan's casting as commentator was a more appropriate fit for the emerging Eurovision audience for SBS: young cosmopolitans with diverse cultural tastes that were not necessarily tied to personal ethnicity. In his book, Mangan (2004, 132) justifies Australian commentary through a cultural interpolation argument that articulates the complex colonial relationship Australia shares with England as well as the common concern about Wogan's growing jingoism:

> while [the BBC's] commentator, the veteran Terry Wogan, was fine for English audiences, a lot of his references were England-specific and, therefore, not really meaningful for an Aussie audience. (Think about it, he was, of course, always going on about how great the English act was and how they should win. This is Australia - even if we're not competing, we still don't want the Poms to win.)

Mangan also hosted preview shows in which panels comprised of entertainers, newsreaders and politicians assessed a portion of entries, as also occurs in some European countries. These were reportedly costly to produce, so were abandoned when SBS returned to the BBC commentary and never reinstated (Egan 2005), although other forms of programming were later developed after the introduction of Pang and Zemiro in 2009. Although Mangan received a handful of praise from my participants, he was not well received at the time; according to Australian Eurovision fan writer Sharleen Wright (2011), fans viewed him as a "poor man's Wogan." One Wogan biography claims, "Viewers were so irate that they jammed the Australian channel SBS's switchboard with complaints, forcing it to run the show again, but this time with Terry fronting it" (Herbert 2016). This may be a biographer's hagiographic hyperbole, but it mirrors the responses to Effie's studio party and the effect was the same. In 2005, SBS returned to the BBC commentary and only sought an Australian team after Wogan's retirement from the Eurovision booth in 2008.

It is entirely possible that Mangan may have been more successful had SBS waited until Wogan's retirement to replace his commentary. The introduction of Sam Pang and Julia Zemiro (heretofore referred to as "Sam and Julia", as they are affectionately known amongst Australian audiences) in 2009 marks a turning point in Australian viewership of Eurovision, arguably beginning the transition into a more participatory form for both the broadcaster and its viewers. SBS invested more effort in its Eurovision programming and, most importantly, were able to develop a stronger and more contemporary branding for Eurovision as something enjoyed by a cosmopolitan Australian audience connected to the contest not just by heritage but as mobile global citizens, which was personified by Sam and Julia as culturally diverse media personalities. Although SBS had been developing its Eurovision resources throughout the 2000s, particularly in its online space, these have proliferated since 2009. Together with their production partner BlinkTV, SBS have produced other content, such as several greatest hits programs and *The Road to Eurovision*, in which Zemiro drove around Europe to interview past and present contestants on her way to Baku in 2012. SBS also host a digital radio station dedicated entirely to playing Eurovision hits in the month of the contest and have cultivated a strong social media presence. SBS were able to gain access to the performers and the backstage even in the era of Mangan, but Sam and Julia were the most successful at establishing themselves as a presence in this space for contestants and viewers alike with the production of backstage footage and interviews. This was facilitated also by the increasing official acknowledgement of the Australian audience at the time, culminating in the provision of a commentator's box to SBS in 2012 at Baku.

The popularity of the team rests predominantly upon fan favourite Zemiro (33.3%); Pang is the third favourite (5.4%), with a ten percentile gap between him and second favourite Wogan (15%), while 3.2% reported that he was their least favourite. In their qualitative comments, participants frequently discuss Sam and Julia's value as a team; some indicated that they would have preferred the team to have been an option in the survey rather than nominating a single commentator.

Julia Zemiro's career encompasses acting, comedy, singing, and hosting on both radio and television. She began developing a fan following from her stint as the host of *RocKwiz*, a music quiz show on SBS, but was already connected to Eurovision in the popular mind via the stage productions, *Eurovision: The Musical* (2003), *Euromax 7: The Musical*

(2004), and *Eürobeat: Almost Eurovision* (2006). Zemiro is praised for her passion, knowledge, and enthusiasm for the contest, as well as her native bilingualism; of French background and educated at a Francophone school in Sydney, she is fluent in both languages of Eurovision—English and French. Her evident love and knowledge of Eurovision itself is identified by fans as key strengths: "Julia just gets Eurovision, she respects it but still has fun with it. I love that she speaks French during the broadcasts, and in interviews. Her music background helps as well, she knows what she's talking about." Zemiro was criticised by a small minority for being too flirtatious with contestants while others admired her charm and confidence in these encounters.

Sam Pang is a comedian who largely worked in radio prior to joining SBS to host *ADbc*, a history quiz show featuring teams of academics and comedians (including Zemiro) that lasted a single season. After Eurovision, Pang continued his radio career and is a series regular on the Channel 10 news comedy quiz show, *Have You Been Paying Attention?* He came to the Eurovision commentary team as the naive foil to Zemiro's expert, an approach worked well in the eyes of most fans who praised the resulting chemistry. Many also observe that the duo—and Pang in particular—improved significantly over their eight years but some focus their criticism on Pang's lack of knowledge about the contest, characterising it in terms of disinterest and disrespect: "Pang hated Eurovision (the music), and it showed." Generally, however, the team of Sam and Julia satisfied most fan expectations regarding humour, knowledge, and passion. Furthermore, as individuals of Asian and French backgrounds respectively they aligned well with the matured multicultural and cosmopolitan brand of SBS in the twenty-first century. Their diversity appears lived in and authentic—it is simply who they are—in contrast to the satire of ethnic femininity presented by Effie or the conscious consumption of cosmopolitanism offered by Mangan.

Citing increasingly busy schedules but with Zemiro later admitting discomfort with Australia's participation as a competitor in the contest (Griffiths 2018), Sam and Julia stepped down before the 2017 contest. The new commentary team comprises Myf Warhurst, a well-known radio presenter, television personality, music buff, and 2016 Australian Eurovision juror, and comedian Joel Creasey, who had appeared on several television programs and comedy festivals but for SBS audiences was perhaps best known for his commentary of the Sydney Mardi Gras that year. Where Sam and Julia personified the cosmopolitan connection, Myf and Joel suggest a more queer- and music industry-oriented approach to the event. Their dynamic does not replicate that of Sam and Julia, but elements are similar. As a long-term fan and former juror, Warhurst is positioned in the passionate expert role previously occupied by Zemiro while, like Pang, Creasey's main role is to provide humour. On balance, the two are easily the least popular commentary team, with Creasey as the least favourite overall (16.7%). Just as Mangan suffered in the eyes (or ears) of Australian viewers for not being Terry Wogan, Myf and Joel suffer for coming on the heels of Sam and Julia. It is reductive, however, to suggest this is the only reason. According to my respondents, Myf and Joel commit several of the severest commentator crimes: they talk over songs, don't demonstrate sufficient or accurate knowledge of the contest, its politics, and its history, and relied upon humour that disrespected the contest. (It is, however, worth noting that audience responses to Myf and Joel improved in 2018.)

Interestingly, fans express some good will toward Warhurst, even as they criticise her poor performance as Eurovision commentator. She is regarded with affection due to her former roles as a radio presenter for youth station Triple J (part of the ABC network) and team captain on popular ABC music quiz show *Spicks and Specks*. These connections to the music industry were integral to her selection as a juror in 2016 and provide fans with a clear rationale for her casting as commentator. As one fan observed, "Myf at least atones [for getting facts wrong] by being amusing and interested in music (and just a 'good egg,' as you might say—I met her once, and that seems to count for something)." Views of Creasey, on the other hand, are less forgiving. The same respondent above adds, "Joel just came across as being an irritant who hadn't done his research." While still criticised by fans, Warhurst acts as a sufficient substitute for beloved presenter Julia Zemiro because she satisfies (albeit imperfectly) fan expectations regarding musical knowledge and genuine appreciation for the song contest. Creasey, on the other hand, was not only less familiar than Warhurst to many viewers but participants felt he lacked passion for the contest. Creasey's television profile is steadily growing, including new projects on commercial networks, but at the time of his casting as Eurovision commentator had not held an established role on television in the same way as his co-presenter at Eurovision, nor even his co-presenters for the Sydney Gay and Lesbian Mardi Gras. He was *known* by many viewers but was not necessarily *familiar*. This could have worked to his advantage, as he could surprise viewers positively, but as was demonstrated in the responses to Warhurst, familiarity can shield commentators from harsher criticism because of affective attachment. Creasey was, in effect, virtually defenseless.

The casting of Creasey is particularly criticised for being "too gay," and an obvious move on the part of SBS to capitalise upon the queer associations of the song contest. The criticisms of Creasey are not levelled at his homosexuality per se, but upon how this element of his humorous approach can be alienating: "Unfortunately SBS didn't realise that this isn't just [a] standard event on the LGBTQI calendar where the generic [camp] humour can be used, it requires knowledge." This respondent emphasises how Eurovision audiences exemplify a contemporary multiculturalism that encompasses all forms of diversity across several generations, from ethnic grandmothers to gay teens. Commentary thus must appeal to both but "Joel's use of generic gay sass left many audience members feeling as if they should not be invited." As Ranier Emig (2014, 277) observes, humour can be utilised to create a sense of community and group identity for those who are in on the joke, but jokes can also be used to exclude. This fan perspective expresses a clear concern with the politics of inclusion and exclusion; this element of the Eurovision ethos has been important to queer fans and they do not necessarily wish to close the contest down for any other groups within the audience because diversity and inclusion are part of the purpose and appeal of the contest.

In many ways, the responses to Creasey are similar to those levelled at Effie's studio party: fans resist the assumptions that are made about them and their identities in Australian Eurovision production. They do not seek to be represented in an obvious or stereotyped way by their commentators beyond a broad sense of diverse Australianness, but they do wish to be guided, entertained, and informed by them in useful, humorous and unobtrusive ways. While more experienced or serious fans may become increasingly ambivalent about the importance of commentary as their own knowledge develops, commentators' knowledge and passion can play an important role in attracting and retaining newer or casual viewers by rendering it accessible and familiar. Commentary thus plays an important role in opening up the event to the uninitiated through the sharing of information, but can also play an important role in confirming the factual knowledge of longer-term viewers. Tone is important here as a "know-it-all" approach can be as alienating as having no information provided at all, and accuracy is paramount. Former SBS Head of Production and Development Denise Eriksen (2017) observes that Sam and Julia resonated with audiences because Pang "knew nothing about Eurovision but was the perfect foil for Julia and could ask the questions others like him needed answering." The lack of knowledge displayed by Warhurst and Creasey combines with the perception that Creasey's humour is exclusionary to "close off" the contest to both new and traditional viewers.

The increased investment of SBS in the Eurovision Song Contest through local commentary, production, and online resources has assisted in further developing the contest into a cult event for its viewers. Local commentary has played an important role in assisting Australian fans to feel a greater sense of belonging to and ownership over the contest by interpolating it into a familiar national context. This is, however, complicated by participation in the contest itself, as evidenced by Zemiro's comments regarding her departure. Australian commentators now risk seeming as patriotic and parochial as their predecessor Terry Wogan, but at the same time must not be perceived as anti-Australian. Furthermore, they must be knowledgable, but not simply relay factual information. They must be humorous, but not disrespectful. Above all, they must mirror and excite the passions of the viewers. While Eurovision fans frequently express an appreciation of tradition, the criticisms of Wogan and the sentiments expressed by Zemiro indicate that perhaps some traditions do have a finite life span, or at least must evolve as the contest itself evolves. One tradition that has undergone an evolution is that of the Eurovision party, to be explored in the following chapter.

References

Andreoni, Helen . 2003. Olive or White? The Colour of Italians in Australia. *Journal of Australian Studies* 27 (77): 81–92.

Ang, Ien, Gay Hawkins, and Lamia Dabboussy. 2008. *The SBS Story: The Challenge of Diversity*. Sydney: UNSW Press.

Carniel, Jessica. 2017. "Fireworks, Feelings, and Fraught Relations at Eurovision 2017." *The Conversation*, May 14. https://theconversation.com/fireworks-feelings-and-fraught-relations-at-eurovision-2017-77396.

Douglas, Kate. 2001. "Increasing the Interactivity: The Eurovision Song Contest and Australian Viewers." *M/C Reviews*, October 25. http://reviews.media-culture.org.au/features/interactive/kdouglas.html.

Egan, John. 2005. "SBS Australia: The What and Why of This Year's ESC Coverage." *ESC Today*, May 13. http://esctoday.com/4467/sbs_australia_the_what_and_why_of_this_year%ef%bf%bds_esc_coverage/.

Emig, Rainer. 2014. "Queer Humor: Gay Comedy Between Camp and Diversity." In *Gender and Humor: Interdisciplinary and International Perspectives*, edited by Delia Chiaro and Raffaella Baccolini, 276–287. New York: Routledge.

Eriksen, Denise. 2017. "The Inside Story of Julia Zemiro and Sam Pang's Eurovision Journey." *The New Daily*, March 3. https://thenewdaily.com.au/entertainment/tv/2017/03/03/julia-zemiro-sam-pang-eurovision/.

Fricker, Karen. 2013. "'It's Just Not Funny Any More': Terry Wogan, Melancholy Britain, and the Eurovision Song Contest." In *Performing the 'New' Europe: Identities, Feelings and Politics in the Eurovision Song Contest*, edited by Karen Fricker and Milija Gluhovic, 53–76. Houndmills: Palgrave Macmillan.

Fricker, Karen, Elena Moreo, and Brian Singleton. 2007. "Part of the Show: The Global Networking of Irish Eurovision Song Contest Fans." In *Performing Global Networks*, edited by Karen Fricker and Ronit Lentin, 139–162. Newcastle upon Tyne: Cambridge Scholars Publishing.

Georgiou, Myria. 2008. "'In the End, Germany Will Always Resort to Hot Pants': Watching Europe Singing, Constructing the Stereotype." *Popular Communication* 6 (3): 141–154.

Griffiths, Neil. 2018. "Julia Zemiro Talks Eurovision Departure: 'It Became Weird To Do'." *The Music*, February 10. http://themusic.com.au/news/all/2018/02/10/julia-zemiro-talks-eurovision-departure-it-became-weird-to-do/.

Hawkins, Gay, and Ien Ang. 2007. "Inventing SBS: Televising the Foreign." *Australian Cultural History* 26: 1–14.

Herbert, Emily. 2016. *Sir Terry Wogan—A Life in Laughter 1938–2016.* London: John Blake Publishing.

Katsabanis, Maria, and Adele Murdolo. 1993. "The World According to Effie." *Lilith: A Feminist History Journal* 8: 71.

Mangan, Des. 2004. *This Is Sweden Calling: Everything You've Ever Wanted to Know About the Eurovision Song Contest but Were Laughing Too Hard to Ask!* Milsons Point, NSW: Random House Australia.

Martin, Adrian. 2001. "Kitsch and Kin." *Sydney Morning Herald*, 23 June, 70, Good Weekend.

McCallum, John. 2004. "Cringe and Strut: Comedy and National Identity in Post-War Australia." In *Because I Tell a Joke or Two: Comedy, Politics and Social Difference*, edited by Stephen Wagg, 200–218. London: Routledge.

McKie, David, and Hilaire Natt. 1996. "An ABC of Australian Sitcoms: British Influences, Middle Class Mores and Boutique Quality." *Continuum* 10 (2): 141–156.

Parker, Heidi M., and Janet S. Fink. 2008. "The Effect of Sport Commentator Framing on Viewer Attitudes." *Sex Roles* 58 (1–2): 116–126.

Sandvoss, Cornel. 2008. "On the Couch with Europe: The Eurovision Song Contest, the European Broadcast Union and Belonging on the Old Continent." *Popular Communication* 6 (3): 190–207.

Turnbull, Sue. 2010. "Missing in Action: On the Invisibility of (Most) Australian Television." *Critical Studies in Television* 5 (1): 111–121.

Wills, Sara. 2004. "When Good Neighbours Become Good Friends: The Australian Embrace of Its Millionth Migrant." *Australian Historical Studies* 36 (124): 332–354.

Wright, Sharleen. 2011. "The Special Relationship: Australia and Its Love of Eurovision." *ESC Insight*, January 26. http://escinsight.com/2011/01/26/the-special-relationship-australia-and-its-love-of-eurovision/.

CHAPTER 4

Part of the Party: Celebrating Eurovision Together

Abstract For many, Eurovision parties are an evolution of the childhood practice of watching the song contest with one's family, but now that "family" is defined by other affective bonds than biological kinship. For Australian fans, the shift from simply an observing nation to a participating nation in the contest has ruptured this tradition because the national audience is split between those who watch the live broadcasts at 5 a.m. and those who watch the delayed broadcasts in the evening. This chapter explores how fans have negotiated this change, including the increasing importance of social media in developing and maintaining a shared national experience of the event while also feeling a sense of connection to other viewers globally.

Keywords Social gatherings · Parties · Celebrations · Social media · Alternative family · Social TV

Prior to 2015, the Eurovision Song Contest was broadcast on the Sunday evening after the event had occurred in Europe, enabling (often boozy) Eurovision parties that night and necessitating the practice of "media lockdown" or "media blackout," wherein fans abstain from news and social media throughout the Sunday in order to keep the contest results a surprise during the evening party. Now that Australia is a competing nation, its viewers are eligible to vote. The results are

J. Carniel, *Understanding the Eurovision Song Contest in Multicultural Australia*,
https://doi.org/10.1007/978-3-030-02315-7_4

therefore not simply a matter of information, but can now be influenced by Australian audiences. Voting must occur during the contest, so the delayed broadcast is no longer sufficient. The contest is now broadcast live to Australia at 5 a.m. (3 a.m. in Western Australia), enabling participation in the official vote, and then replayed in its traditional 7.30 p.m. time slot.[1] More than half of the participants (55%) report watching the 5 a.m. broadcast, 20% wait until the evening replay, and a handful watch both broadcasts. An additional 9% specified that they watch the contest on the Eurovision.tv stream, indicating a live view as well as ambivalent feelings about the role of local commentary and production. Many fans are pleased that it is now much easier to access a live broadcast, as in the past they would have to seek out alternative (and sometimes pirated) sources on the internet if they wanted a live rather than delayed viewing. Changes to the broadcasting have in turn changed the way that Australians celebrate Eurovision together.

Eurovision is an example of ritualised event television that works to gather family and friends around the modern hearth of the television. Many fans in Australia and around the world celebrate this gathering as a Eurovision viewing party. Generally, this is a party for the grand final event wherein friends and family gather to share the experience, although some Australian fans focus instead on the semi-finals, as the delayed broadcast for these events usually occurs on the more party-friendly Friday and Saturday nights. Some fans may also hold "preview parties" that screen the official videos of each selected performer, as available online via YouTube and the official Eurovision website, rather than focusing on broadcast events, thus enables a celebration and shared experience of the music that bypasses some of the logistical challenges of the event for Australian viewers. Eurovision parties feature food and drinks (often themed), costumes, and games. Eurovision parties became a more prominent tradition for Australian fans during the 2000s and 2010s. In its early days in the 1980s and 1990s, fans were more likely to share the event with family rather than friends or, if even watching with friends, it was less likely to resemble a party as such. SBS promote Eurovision as a social

[1] This describes the process for the grand final only. The semi-finals are held live on Tuesday and Thursday of the finals week, thus screening live on Wednesday and Friday mornings in Australia, but the delayed broadcasts are screened on Friday and Saturday nights. This scheduling was altered in 2018 to better generate contemporaneous support for Jessica Mauboy, but many fans expressed displeasure with this change via social media.

occasion in its advertising, online resources as well as social media campaigns and competitions that ask fans to submit images of their parties. As Australia's place within the competition has changed, so too have the celebratory traditions surrounding it. Although the Sunday night grand final party may not have been ideal for those going to work the next morning, for many the new 5 a.m. broadcast hour presents the greater challenge. Australian fan practices have had to adapt accordingly.

The idea of television as the modern hearth derives from David Morley's *Family Television: Cultural Power and Domestic Leisure* (1986), wherein he describes the television as the "family hearth" of the modern household. Writing in the mid-1980s, Morley also emphasises how other technologies, such as video, teletext, and video games were beginning to change the role of the television in the domestic setting and, as a consequence, the household relations around it. Domestic television viewing in the twenty-first century is certainly a very different affair to that described by Morley. Not only are households now likely to have more than one television set, they also have multiple other screens, such as computers, tablets, and mobile telephones, that are all capable of streaming television programs and other media. Some streaming services are appended to broadcast services, such as SBS On Demand, providing an online access point to programming being broadcast at a particular time as well as enabling a "catch up" service. Generally, however, streaming enables greater detachment from scheduling than broadcast television, as viewers can choose what to watch when they want to watch it. On the one hand, television as a social household ritual is potentially endangered as each member can watch what they choose in isolation. On the other hand, opportunities to watch programs together can become more valuable and the events themselves more important because of their rarity. Streaming television may make the ritual schedule more flexible to a household's needs, but significant broadcast events can not only bring the household to the modern hearth of the flat screen television, but also forge connections with other hearths around the nation and the world.

In their seminal work on event television, Daniel Dayan and Elihu Katz (1988) use Durkheim's idea of civil religion to unpack how television viewers develop active responses to broadcast events, which further underscore the events' significance and cultural value as a form of modern ritual. The specialness of such events prompts a more active and social form of television spectatorship than regular viewing. While they emphasise that modern television events can be more irregular than Durkheim supposed, such as royal weddings, particular sub-genres of televised events, specifically contests such as the Olympics or Eurovision do occur with greater regularity. These events interrupt quotidian programming, but are in themselves regular events, occurring annually or every four years at a specific time in that year. Whether regular, like Eurovision and the Olympics, or irregular, like a royal wedding, these media events are special occasions that prompt different viewer behaviours in part due to a renegotiation of public and private space: "Public space invades the home. People no longer undress to watch television; they may actually dress up. The event becomes the occasion for a spectatorial contract. Friends are invited to watch the show" (Dayan and Katz 1988, 162). This can be readily seen in an event like Eurovision, which prompts festive social gatherings and costumes. Social media further blurs this line between public and private spaces so that the event shared at home is shared experience in not just a symbolic sense but in more tangible ways, such as sharing observations via text, Twitter, or Facebook. There is also a visual element to this shared experience and blurring of spaces as people post images of their parties or bleary-eyed selfies of early morning viewers on platforms like Instagram.

Eurovision parties occur globally, but few scholars of its fandom have engaged in depth with the phenomenon. Brian Singleton et al. (2007, 16) touch upon Eurovision parties in their study of Irish fans, describing them as "elaborative and performative" events involving food, games and banter. Myria Georgiou's (2008) study of how stereotypes are deployed as expressions of everyday racism and xenophobia in public debates about Eurovision also engages briefly with the celebratory habits of young university students in her focus group. Eurovision parties, Georgiou (2008, 149) observes, "provide one kind of platform for consuming the event live and communally", while also providing "a terrain where the distant becomes familiar, or appropriated as an exotic, non-threatening other." Social media now provides another important live, communal platform, but both Facebook and Twitter were fairly new at the time that Georgiou was writing. Accordingly, her participants place greater emphasis upon co-presence in the living room than other forms of community, although one participant does mention setting up a "live video phone link" with her family during a year abroad. Finally, Dafna Lemish (2004) considers Eurovision parties in her study of gay Israeli fans. Again, the discussion is brief but revealing. It highlights the importance of parties for providing community and escapism for a minority group; again, this is also supported by Singleton et al.'s findings amongst queer Irish fans. Furthermore, her participants make an interesting connection between Eurovision and ideas of ritualistic and religious group celebrations that recurs in my own findings.

For Lemish's (2004, 51) participants, the ESC is a "special social event," which one likens to "Passover for the homos." The simile reveals the importance of Eurovision as a time to gather together in celebration, to share food, and to share an important cultural text. Importantly, Lemish highlights the different kinds of "families" that are involved in Passover and Eurovision. Traditionally, Passover is celebrated with biological family, while Eurovision is an event shared with "alternative 'families' of close friends" and functions as a familial celebration for that group. Australian fan Harry uses a similar religious metaphor for Eurovision and its fandom, albeit derived from Christianity. He refers to the finals week as "Holy Week" to indicate a time of ritual importance in his personal calendar. This metaphor, he admits, offended some Catholic friends, but he presents a clear sociological rationale for its usefulness based upon undergraduate studies of world religions and cults: "[Eurovision] has its own language, it has its own rituals, and hell, it even has its own shrine." He refers here to an exhibit set up in a Stockholm museum during the 2016 Eurovision where he was photographed with the golden shoes from Herreys' performance of "Diggiloo Diggiley" (Sweden 1984). "These are artefacts in the same way as, I suppose, the Catholics have their relics, I've got mine."

The Irish, British and Israeli contexts of fandom or enthusiasm and parties are, of course, different to Australia. One important difference is that the UK, Ireland and Israel are less affected by time differences; for example, Israel and Ukraine are in different continents but share the same time zone. This is one less factor to which fans must adjust in organising their festivities. As the live broadcast is part of these cultures as participating nations of relatively long standing (the UK has participated since 1957 and Israel joined in 1973), practices such as media blackouts or avoiding spoilers are less important than it is for the delayed broadcast culture in Australia. Even though many Australians do watch the live broadcast, many still do not. Fan practices around viewing have necessarily hybridised to accommodate both options. Despite these differences, the British, Irish, Israeli, and Australian studies combine to highlight key themes in global Eurovision fan culture: community, sociality, and family (either biological or alternative). Eurovision is a time to be with others, either in person or in the virtual space, and to celebrate shared joy in the event, even as the meaning ascribed to this event may differ from individual to individual.

Seventeen per cent of the Australian survey participants view the event with their family and 19% view with a combination of both friends and family. One quarter of participants watch with friends only, while 10% watch with their partners only. Twenty per cent of participants reported that they generally watch the event alone. Notably, the majority (81%) of those solo viewers also reported that they participated in social media as part of their fan and viewing practices; one even specified that they do not really watch it alone, as they are accompanied by their fellow fans on Twitter. Some of these participants also indicated that they watch both the live 5 a.m. telecast and the repeat broadcast in the evening, watching the former alone and the latter with friends. Those who watch with family are more likely to view the delayed broadcast, particularly when small children or the elderly are involved, which in turn affects the degree to which they can participate in elements such as the voting. Who you watch it with and the reason why you watch it in turn influences the way in which the event is celebrated. For some fans, Eurovision has traditionally meant a quiet Sunday evening with family (acknowledging, of course, that families can also be fairly raucous!); for others, it is a boisterous evening featuring drinks with friends.

From its outset, the contest's aim was to produce wholesome viewing that would bring both Europe and its families together. In his history of Eurovision, Chris West (2017, 10–11) paints an evocative picture of the family as television audience in the 1950s when the contest first emerged:

> The *Grand-Prix* was aimed at a middle-class family audience because the EBU thought that would attract the most viewers. Given the profile of TV ownership at the time, they were probably right. Oma, Mama, Papa and the younger kids would all watch it; if their stroppy seventeen-year-old wanted to sneak up to their room and listen to Alan Freed on their new transistor radio, well, let them. They'd soon grow out of it.

This image is perpetuated in contemporary industry perspectives of the contest. As Swedish television producer, former Eurovision contestant, and Contest Producer (2016–2018) Christer Björkman states, "The aim is to get the whole family to sit down together - to create a social phenomenon like Christmas with songs for everybody" (EBU News 2015).

The small percentage of participants in this study who began watching Eurovision in the 1960s and 1970s during their childhoods abroad do associate it explicitly with the idea of family viewing: "[it's] what the family did." Those who began viewing it in the early days of the SBS broadcast in the 1980s are also more likely to associate it with family viewing and the beginning of a family tradition; again: "It was something our family did each year." Some attach the event to the tastes, idiosyncrasies, or origins of a particular parent—"My Dad is English - he has always followed Eurovision. It was all about bonding with my Dad"; "Mum is from Belfast and wanted to share something they did back in the day"—while for others it simply emerged as a new family tradition for those who were already watching television together on a Sunday evening. For many fans, this childhood viewing provides the foundations for lifelong fandom (Sandvoss 2008).

Eurovision fandom is also intergenerational. One fan who has been watching since the late 1990s enjoys Eurovision as a "great event for adults and kids," seeing it as an "opportunity to discuss music and learn about other countries with the kids." Another fan who has been watching since the mid-1990s after being encouraged to do so by her own mother now takes turns in hosting duties with friends from a variety of Scandinavian nationalities. Their children "use a chalkboard to tally favourite songs as they appear in the broadcast." In the first example, watching Eurovision is an entertaining and educational experience shared by the nuclear family, much as it was first conceived to be. The second example, however, demonstrates how ideas of "family" and event television have broadened in the twenty-first century beyond the biological. It is something that draws not just an individual family but many families bound by friendship to create a communal event shared in the physical space of the domestic home, connected also to ideas of hospitality as something that is shared equitably.

Ideas of "family" play an important role in understanding the Eurovision audience globally and in Australia but this concept must necessarily be broadened to include affectively bound social units that are not based upon biological kinship. Singleton, Fricker and Moreo (2007) emphasise Eurovision as a form of family entertainment and credit this characteristic with its success, but they also highlight that "family" in the twenty-first century is no longer defined purely in biological terms. Eurovision can play an important role in creating and defining these new, alternative familial relations. As a result, "family entertainment also has an alternative family, a queer family" (Singleton et al. 2007, 11–12), a theme that emerges also in Lemish's Israeli study. For many, "urban" or "alternative" families are the those with whom the experience of watching the Eurovision Song Contest is shared, particularly in the 2000s.

The concept of alternative families is, of course, not limited to queer culture as many people of diverse identities develop networks of "urban family" through friendships and the experience of share housing in their university years and early professional lives. Despite the ubiquity of this domestic arrangement, scholarship on share housing is limited and tends to be focused upon demographical implications, such as housing affordability (Clark et al. 2017). It characterises share housing predominantly as transitional and defines the household in opposition to the family home as "non-family" living (McNamara and Connell 2007). Such perspectives overlook the important affective dimensions of the share house built upon various shared experiences, as well as other forms of family or home making that can occur in social groups, even if they do not cohabit. Socio-economic factors are also important to consider in these arrangements as they can affect accessibility to resources, which in turn can even affect how and why individuals engage with cultural texts such as Eurovision. For example, Joanna first started watching Eurovision because many of her share house members were "unemployed, underemployed, working weird hours," which gave them "a bit too much time to sit around on the internet." It was through passing time in this manner that she and her flatmates encountered Moldova's "Epic Sax Guy" (Sergey Stepanov from SunStroke Project in 2010 and 2017) during the height of his viral fame. Joanna's story is linked to her specific socio-economic circumstances and domestic arrangements at the time, but more importantly it illustrates how cohabitation leads to shared experiences that in turn create or foster existing affective connections. Kara, for example, describes her living arrangement with Drew and their third flatmate David as "a house family." While she does not identify as a Eurovision fan and claims that she would not watch it on her own, she nevertheless values Eurovision as an important shared activity that bonds the household together; this mirrors the original intentions of the contest but in a new domestic milieu.

The division between biological and alternative families is by no means absolute. Eurovision also provides some fans with an opportunity to bring together these groups for a shared experience. Ryan describes the evolution of his preview parties, which he has held in preference to finals parties as he has frequently attended the event as a member of the fan press corps: "I think they all just sort of thought, 'Well, you know, this is another thing that Ryan's into,' or whatever…I think it's sort of evolved from there to them being a little bit interested. Now it's a must-not-miss for family and friends." Here, biological and alternative kinship groups are brought together, united by their relationship with Ryan and their growing appreciation for the contest. Ana's narrative is similar, although her viewing began with her Greek family already having familiarity with the contest and its history. She first viewed the contest with her grandfather and brother, but now she considers it as "a chance for me to hang out with my friends and family." She also describes Eurovision in general has having "a family sort of feel to it," not just because of the family-friendly content but the opportunity it offers for togetherness.

As these examples demonstrate, one perceptible trend amongst most Australian Eurovision fans is the value of the event as a reason to bring people together. While the Eurovision event itself is appreciated as a musical spectacle, its value as a social event leads some fans to prioritise the opportunity to watch it with family and friends over the importance of watching it live. While interviewee Chris was not undeterred from watching it live, he nevertheless was hopeful that friends would join him and his wife in the morning viewing as the social element makes it "real"; that is, sharing the event makes it something more than simply a television program, augmenting the ethos of togetherness that many associate with the contest.

Gatherings or parties are important as they secure Eurovision's place within familial or social rituals. What occurs at these parties helps to create new memories for that group. For example, Singleton et al. found that at-home commentary is an important feature that works to underscore the contest's social and interactive quality. Attendees "engage in their own contest to provide the wittiest, bitchiest and most knowledgeable commentary on the singers and their acts" (Singleton et al. 2007, 16). This commentary also functions to help ritualise Eurovision amongst the family unit, whether biological or alternative, as past viewing experiences of the group come to form an important part of the overall Eurovision fan narrative. Eurovision is therefore not just about what you see on the screen, but about memories of shared experiences at these events. Arguably, social media extends the reach of this at-home commentary, therefore potentially extending the shared experience of the event to the imagined community of the nation—and the imagined community of Eurovision fandom.

Historically, the delayed broadcast meant that Australian fans were only starting to celebrate half a day after European fans had finished their parties. Many Australian fans would engage in a media blackout in order to preserve a sense of live-ness to the events they were watching, which in turn enabled participation in games, voting, and betting that would otherwise be spoilt by knowing the contest's outcome. Furthermore, these parties occurred on a Sunday evening during late autumn rather than a Saturday evening in spring, as it is in Europe; the chill and the need to go to work or school the next day caused some to be more circumspect about their celebrations. These factors have affected the way that this tradition has evolved, and continue to be a consideration with the advent of Australian participation.

Australian participation in the contest has increased general awareness of and interest in Eurovision. Older fans are somewhat relieved that they "don't have to explain as much what a Eurovision party is anymore," but entrance into the song contest has disrupted the logistics of the Eurovision party tradition. This is not to say they do not still occur. As interviewee Simon observes, "that's sort of what we do, you know. Have a Eurovision party." Sixty per cent of participants still either host or attend a Eurovision party, but how and when these parties occur has changed.

SBS has used the tradition of Eurovision parties as part of its promotion strategy for contest over the years. Australian fans have been an increasing presence at the live Eurovision event, even in the years before official participation, but for most viewers and for SBS the emphasis has largely been upon the parties than upon attending the live event. As a broadcaster, SBS is, of course, primarily invested in the success of the televisual event. Its promotion of fan parties demonstrates a clear understanding of how fostering sociality around a program can bolster audiences. SBS's Eurovision website features party guides and resources, which have also been promoted in other publications, such as magazines and email newsletters (SBS, n.d.). Parties also feature prominently in their social media feeds. In one promotional campaign fans were encouraged to use the designated #SBSEurovision hashtag to share their pictures of their party with fellow fans around Australia, entering the draw for a prize in the process.

Eurovision is a useful cross-platforming opportunity that enables the creation and promotion of new and existing content. As food often plays a central role in these parties, cross-platforming with food-related media on SBS provides a useful example. SBS is a strong contributor to Australian food culture, but with a clear multicultural, cosmopolitan focus that aligns with the network's brand. Food-related programming, such as cooking shows and documentaries, feature heavily on the SBS multichannels, one of which is now its dedicated Food Network, the only free-to-air channel in Australia focused on food culture and cooking. The SBS website also includes a dedicated food section that, together with the Food Network, replaced their monthly food magazine, *Feast*, which folded in 2015. In recent years, this has featured Eurovision-related content, which is in turn linked to other SBS content and platforms. For example, Adam Liaw, a former *Masterchef Australia* winner (screened on commercial Network 10) who has since moved into the SBS talent stable, created a series of Eurovision-related recipes in 2017 (SBS 2017). Liaw, a self-confessed Eurovision fan, also curated the @SBSAustralia Twitter account during the 2017 first semi and grand finals. While his recipe collections are Eurovision-themed and linked within the Eurovision pages, the original posts and all other food-related Eurovision posts are hosted within the Food Network pages of the SBS website, not the Eurovision pages. The Eurovision pages did feature a humorous article listing reasons why Liaw ought to be considered for the role of commentator (Celjo 2017), using this also as a platform for promoting his cooking and travel program, *Destination Flavour*, for which a special Eurovision-themed episode was produced in 2018.

Participants frequently describe the food prepared for their parties in general terms as "European" or "European-style" food, which can range from delicatessen platters of cheese and cured meats, to specific national dishes like spanakopita or kofte, or even a simple pizza (delivered is fine). The introduction of the live telecast has impacted upon Eurovision food culture in Australia as many traditional items are not palatable (Liaw's recipe for German sausages with homemade beer mustard, for example, may be too rich for breakfast fare) or even practical in the early hours of the morning (most pizza places are not open at that time). For some, Eurovision food culture has transitioned into a feast of European breakfast items instead, but others express difficulty in adjusting to the creation of a new early morning tradition. As one participant reports, "we're less likely to host/attend a Eurovision party, as 5am on a weekday [for semi-finals] is a tough ask for most. I know people who have evolved to having a 'Eurovision final breakfast', but that's much less fun than getting pizzas, I feel." Another participant also feels that even when they do occur "parties are less party-like" at 5 a.m. The decline in party atmosphere may be because drinking games have been a casualty in the transition from delayed to live telecasts, although they do still feature for those at evening viewing parties. Some have initiated Eurovision slumber parties as an alternative. While these are still fun, one slumber party host adds, "if I'm being perfectly honest, I miss the evening parties!" The atmosphere is often different when attendees are just waking up in the early morning in comparison to a party held at the end of a day, although Chris describes arriving the morning after a slumber party attempt by friends: "You could tell which ones had really enjoyed the night before and were still going." Those who watch both broadcasts are more likely to preserve the party tradition by watching the live broadcast in the morning, resting throughout the day, then hosting or attending a party in the evening. "The experiences are really different," one participant explains of this practice, "and I get a lot from both."

For live-view fans who used to engage in media blackouts, the 5 a.m. broadcast has relieved some of the pressure around the event as they no longer need to avoid media outlets, including social media, throughout the day. "I do like watching the grand final live because no matter how careful you are, inevitably someone leaks it to you, you know, between when it actually happens and when SBS broadcasts it at night," Joanna explains. For those who conserve their energies for the evening broadcast, it is even more difficult to avoid news of the event's outcome as other media outlets—not just SBS—are more likely to report on the event now that Australia participates. According to Tracy, "The ABC are really bad for spoilers!" Not knowing the results while viewing is as important as voting for many fans as gatherings can involve polls or sweepstakes. According to one participant, these voting games are "the most important aspect" of his parties. Each attendee must "predict the winner and the loser", which obviously cannot occur if the results are known. Watching the event in real-time has also enabled some Australian fans to connect their parties transnationally, using voting as part of that. One participant now hosts a breakfast party at the live telecast during which they connect with overseas friends in real time: "We have breakfast with themed food, a few costumes but a religious set of voting (that I insist every person attending do). We tally our party's vote and then share those over Skype with friends' parties in the UK."

For more avid fans, the final results do not define the contest entirely nor do they limit opportunities for celebrations and gatherings. Fan groups, such as the Australian branch of the OGAE (the official Eurovision fan organisation that operates globally with national chapters), often host events throughout the year, including preview parties before the event and other ad hoc celebrations featuring Eurovision music or trivia. Immediately following the contest, much fan activity centres upon dissecting the contest and evaluating the accuracy or fairness of the results. Two fans have integrated this into a party tradition that somewhat mirrors the Skype party described above but that does not rely upon the live event nor the delayed broadcast as its lynchpin. This pair is fortunate to be able to attend the live event each year, so after their return they host a "rerun" to be enjoyed with their friends at home, but which also connects to a network of other rerun parties around the world. In addition to traditional Eurovision party elements, such as costumes and themed catering, the reruns involve "international jury voting" wherein they call for results from other parties held within their global network of Eurovision fans, and report their own in return. The rerun creates an alternate outcome for the contest that does not rely upon its live-ness, focusing instead on the real-time interaction between the international "juries" and the party attendees, as well as the musical knowledge and tastes of these participants.

For many, the live "party" now occurs in the online space as this is an easy way to negotiate communal enjoyment of the event at its party-unfriendly time. One participant describes their parties (using this term specifically) as "online mostly via text and voice chat (lots of emojis) and ROWDY [sic]." As illustrated by previous examples, Skype is a useful tool for directly connecting groups of friends and family, as is SMS texting, while social media platforms are more useful for accessing networks beyond personal acquaintance. Although some public groups are available, including an official SBS Eurovision fan group started in 2018 (distinct from the official SBS Eurovision page, which is largely a platform for communicating official news rather than fan-driven content and discussion), Facebook is generally used for conversations among friends. Some fans report that the week of Eurovision is the only time they interact meaningfully with some Facebook friends who they are connected to through a shared appreciation of the contest. By contrast, Twitter is mentioned frequently by participants as a preferred means for synchronous public commentary on the event: "I think it's a good way of interacting with people while it's still 5am," explains Chris.

Social media platforms like Facebook and Twitter are easy means of sharing information and commentary, but they are also a means of building and experiencing a sense of community around a common interest or, in the case of live tweeting and television viewing, a common event. Live tweeting has emerged as an important part of experiencing a variety of events in the virtual space, from television (Ji and Raney 2015;

Pittman and Tefertiller 2015) to academic conferences (Groves 2016) to newsworthy global events (Sullivan 2014; Hounshell 2011). It facilitates the sharing of information but the dialogic nature of Twitter also enables a sense of community-building through sustained interaction and communication. Ruth Deller (2011, 223) identifies this communal aspect of Twitter as amongst the most important dimensions of the platform, with hashtags and trends being "markers of communality" that assist viewers in constructing a shared space. These, according to Qihao Ji and Arthur A. Raney (2015, 226–227) "signal membership in the same club, shared fandom; they invite sociability." Twitter in particular appeals to fans because of its "sense of immediacy, presence albeit a virtual one, and connection" (Bury 2017, 637) and, unlike the "walled garden" of Facebook (Bury 2017, 638), it enables and encourages engagement with other fans outside one's immediate personal acquaintance. The boundaries of Twitter conversations are defined by the use of hashtags that construct a "virtual loungeroom" (Highfield et al. 2013). Social media means that the conversation about and, by extension, the party for Eurovision is occurring on a broader level that both is and is not defined by geographical space and time.

While some may focus their conversations on their followers and those they follow themselves, others use hashtags to seek out the comments of other audience members who might not otherwise be known to them. Global conversations about Eurovision centre on #Eurovision, while #SBSEurovision helps signify a local view. During Eurovision finals weekend in 2015, over 134,000 tweets used the official #SBSEurovision hashtag. This SBS hashtag performs the important function of localising the broadcast for both the broadcaster and its audience (Highfield et al. 2013); similar hashtags are used by other broadcasters, such as #BBCEurovision, #RTEEurovision, and #EurovisionTVE. Broadcasters use a variety of tactics to encourage use of the local hashtag, such as linking them to competitions. SBS's delayed broadcast includes a Twitter feed in the bottom corner of the screen that features tweets using #SBSEurovision. Getting a witty quip into this feed is a highly prized achievement for fans. "I think I've succeeded once with a really bad pun," Joanna recalled. "That country act from the Netherlands called The Common Linnets a few years ago? Their performance was, like, highway themed and I made some absolutely terrible pun about it being a middle of the road performance. Of all the tweets I could get on TV, it would have to be a really terrible pun."

Localised hashtags are useful for both the audience and the broadcaster. As Tim Highfield et al. (2013) observe in their case study of

Eurovision-related hashtags, localised hashtags enable broadcasters to track audience responses, thus working as a form of market research. For example, tweets using the local hashtag do not comment simply on the contest, but also on the commentary. This allows producers to provide feedback to the commentators on their performance. For Australian fans, the localised #SBSEurovision hashtag allows them to construct a fan space that accommodates the specificities of their shared time and place in contrast to European or global fans using #Eurovision. Fans may also use this platform to correct information provided by the commentators; these corrections are useful for both the commentators and the audience. In the past, the #SBSEurovision hashtag was also helpful for Australian viewers to avoid spoilers from the #Eurovision hashtag while watching the delayed broadcast, although this has dissipated somewhat as many 5 a.m. viewers use both the local and global hashtags. In addition to its practical uses, the local hashtag provides audiences with an opportunity for interaction with each other in a public forum. This interaction can facilitate a sense of belonging to a national audience and, through this, a national community. Interviewee Simon notes the national conversation on Twitter as a specific element he missed while watching Eurovision in Scotland in 2015:

> All the people I know in Australia watch Eurovision on Twitter and it's just sort of very funny on Australian Twitter and I just didn't get as much of that in UK. Probably because when you look up the Eurovision hashtag, it's more Europeans who are speaking different languages, so it can be harder to follow. Whereas when you look at the SBS Eurovision hashtag, it's only used in Australia and it's on Sunday night when Eurovision is done, so it's a bit less, it's a bit easier to follow and to connect with people. So I did miss that a little bit, although there were definitely people I knew who watched it live back in Australia.

Fans often use a combination of both hashtags, depending upon how they wish to access the conversation occurring, allowing them to either shift between or to merge their virtual loungerooms of the global #Eurovision and the local #SBSEurovision. SBS chose to abandon #SBSEurovision in 2018 in favour of #Eurovision in order to be part of a global conversation. This complicated SBS's own use of #SBSEurovision as a production element of the delayed broadcast, in which fan tweets are posted on the screen, as fans watching the delayed broadcast complained of spoilers and even tweets that were clearly posted by people outside of Australia who were not watching the SBS broadcast. Although SBS attempted to steer fans toward their specially-created Facebook group for localised conversation, many on Twitter used the local hashtag defiantly (although some were simply not aware of the change). By persisting with #SBSEurovision, Australian viewers expressed a desire and need to conduct their own national conversation that addressed the idiosyncrasies of Australian time, place, and commentary, even as they valued participation in a global event.

While some fans might express a sense of loss or nostalgia for the old way of celebrating Eurovision, few express a sense of loneliness about watching the live broadcast, even if they are watching it alone in their physical space. The phenomenon of synchronous online conversations about television programs is known as "social TV" (Buschow et al. 2014, 129). Highfield et al. (2013) use the analogy of the "virtual lounge room" to signify the "communal space where audience members can come together to discuss and debate, in real time, their responses to what they are watching on the television screen," much as one would in one's own living room. As suggested above, hashtags play an important role in helping viewers to identify one another and to define the space—the walls of the lounge room, so to speak—in which the conversation occurs. Social media means that the conversation about and, by extension, the party for Eurovision is occurring on a broader level that both is and is not defined by geographical space and time. Individuals are gathered around their own "modern hearths," but all these hearths may be installed in the same "virtual lounge room."

Watching Eurovision is not a static, passive, or uniform act for its fans. While they value the text itself, they also clearly value its capacity for interaction and sociality. Australian fans of Eurovision are highly aware of the song contest as a constructed television event and yet it is something that they value highly because it offers them a sense of community and connection through the means of fun and entertainment. While Australian participation in the contest and the subsequent 5 a.m. live broadcast has disrupted some Eurovision traditions for Australian fans, it has also offered an opportunity to broaden and even deepen the communal experience of Eurovision from the physical and into virtual spaces of fan interaction and community. The challenges of time and geography, rather than being impediments, work to make this belonging and community even more meaningful. As shown here, Australian fans use a variety of in-person and online strategies to enjoy the event and to cultivate a sense of belonging to its fan communities on local, national, and international levels.

References

Bury, Rhiannon. 2017. "Technology, Fandom and Community in the Second Media Age." *Convergence: The International Journal of Research into New Media Technologies* 23 (6): 627–642.

Buschow, Christopher, Beate Schneider, and Simon Ueberheide. 2014. "Tweeting Television: Exploring Communication Activities on Twitter While Watching TV." *Communications—European Journal of Communication Research* 39 (2): 129–149.

Celjo, Farah. 2017. "19 Reasons Adam Liaw Is (so) Eurovision Ready." *SBS.com.au*, March 9. https://www.sbs.com.au/food/article/2017/03/06/19-reasons-adam-liaw-so-eurovision-ready.

Clark, Vicky, Keith Tuffin, Karen Frewin, and Natilene Bowker. 2017. "Shared Housing Among Young Adults: Avoiding Complications in Domestic Relationships." *Journal of Youth Studies* 20 (9): 1191–1207.

Dayan, Daniel, and Elihu Katz. 1988. "Articulating Consensus: The Ritual and Rhetoric of Media Events." In *Dukheimian Sociology: Cultural Studies*, edited by Jeffrey C. Alexander, 161–186. Cambridge: Cambridge University Press.

Deller, Ruth. 2011. "Twittering On: Audience Research and Participation Using Twitter." *Participations: Journal of Audience and Reception Studies* 8 (1): 216–245.

EBU News. 2015. "Eurovision Song Contest Workshop Focuses on How to Achieve Success and Build Audiences." *EBU.ch*, September 14. https://www.ebu.ch/news/2015/09/eurovision-song-contest-workshop.

Georgiou, Myria. 2008. "'In the End, Germany Will Always Resort to Hot Pants': Watching Europe Singing, Constructing the Stereotype." *Popular Communication* 6 (3): 141–154.

Groves, Trish. 2016. "Tweeting and Rule Breaking at Conferences." *British Medical Journal* 353: 3556.

Highfield, Tim, Stephen Harrington, and Axel Bruns. 2013. "Twitter as a Technology of Audiencing Fandom: The #Eurovision Phenomenon." *Information, Communication & Society* 16 (3): 315–339.

Hounshell, Blake. 2011. "The Revolution Will Be Tweeted." *Foreign Policy*, no. 187: 20.

Ji, Qihao, and Arthur A. Raney. 2015. "Morally Judging Entertainment: A Case Study of Live Tweeting During *Downton Abbey*." *Media Psychology* 18 (2): 221–242.

Lemish, Dafna. 2004. "'My Kind of Campfire': The Eurovision Song Contest and Israeli Gay Men." *Popular Communication* 2 (1): 41–63.

McNamara, Sophie, and John Connell. 2007. "Homeward Bound? Searching for Home in Inner Sydney's Share Houses." *Australian Geographer* 38 (1): 71–91.

Morley, David. 1986. *Family Television: Cultural Power and Domestic Leisure.* London: Comedia Publications Limited.

Pittman, Matthew, and Alec C. Tefertiller. 2015. "With or Without You: Connected Viewing and Co-viewing Twitter Activity for Traditional Appointment and Asynchronous Broadcast Television Models." *First Monday* 20 (7). http://www.ojphi.org/ojs/index.php/fm/article/view/5935/4663.

Sandvoss, Cornel. 2008. On the Couch with Europe: The Eurovision Song Contest, the European Broadcast Union and Belonging on the Old Continent. *Popular Communication* 6 (3): 190–207.

SBS. 2017. "Adam's Eurovision Bites Recipes and Adam's Eurovision Bites Food." Accessed June 29. https://www.sbs.com.au/food/program/adam-liaws-eurovision-bites.

SBS. n.d. "Party Guide." *SBS.* https://www.sbs.com.au/programs/eurovision/party.

Singleton, Brian, Karen Fricker, and Elena Moreo. 2007. "Performing the Queer Network: Fans and Families at the Eurovision Song Contest." *SQS–Suomen Queer-tutkimuksen Seuran lehti* 2 (2): 12–24.

Sullivan, Rachel. 2014. "Live-Tweeting Terror: A Rhetorical Analysis of @ HSMPress_ Twitter Updates During the 2013 Nairobi Hostage Crisis." *Critical Studies on Terrorism* 7 (3): 422–433.

West, Chris. 2017. *Eurovision: A History of Modern Europe Through the World's Greatest Song Contest.* London: Melville House.

CHAPTER 5

"We're a Nation of Proud Multicultural Dags": Multiculturalism and Viewer Identities

Abstract The dominant narrative for explaining the popularity of the Eurovision Song Contest in Australia is that it signifies the ongoing connections to Europe built from post-World War II mass migration. While this is certainly the initial foundation for the contest's popularity, this "European connection narrative" relies upon an image of "multicultural" Australia defined by historical whiteness and Europeanness. This narrative dominates official commentary of the event but also appears in fan perceptions of the contest's Australian popularity—yet few fans identify ethnic identity or heritage as their defining motivation for watching the contest. Ethnicity emerges as part of the texture of Australian Eurovision fandom, but it is far from its defining feature. Nevertheless, Australian fans feel strongly connected to ideas of multiculturalism and, through this, the world.

Keywords Multiculturalism · Australian immigration · Transnationalism · Pop cosmopolitanism

Announcing Australia's return to Eurovision in 2016, EBU Reference Group Chair Dr. Frank Dieter Freiling described the contest as "a way for many Australians to reconnect with their European roots" (quoted in Jordan 2015b). We see this mirrored in participant responses, such as: "I know a lot of people watch it because their families come from

J. Carniel, *Understanding the Eurovision Song Contest in Multicultural Australia*,
https://doi.org/10.1007/978-3-030-02315-7_5

Europe and it's a way of staying connected." Such comments paint a picture of a predominantly European Australian audience that was not reflected in the demographic data of my sample. Of the participants interviewed and surveyed for this study, three interviewees were migrants (from Sweden, the UK, and Greece, although the latter migrated as an infant), four identified non-Australian and/or non-Anglo Celtic heritage, and 20% of all survey participants also identified as having migrant heritage. While these participants were overwhelmingly from a variety of (sometimes mixed) European backgrounds, it is noteworthy that 22% of those respondents identified an Asian background. What these results suggest is that while ideas of heritage and cultural diversity are undoubtedly a part of the Australian understanding of Eurovision, they are not as defined by notions of Australia's "Europeanness" as the quotes above might suggest, nor are they necessarily primary motivators for individual fans to watch the contest. Nevertheless, the "European connection" narrative persists.

Themes of multiculturalism or migrant/ethnic heritage feature in 16% of qualitative survey responses to the question, "Why do you think Australians in general love watching Eurovision?", which asked about their perception of its broader popularity as distinct from the respondent's personal reasons for watching. This was the third most common perception after themes focused on Australians as fun-seeking and as having a camp or kitsch cultural sensibility. There is some overlap in the reasons why individuals themselves enjoy Eurovision, but ethnic heritage or connection is not high amongst these; participants prioritised fun, the contest's camp or kitsch factor, and the music, while ethnic belonging or identification ranked sixth. When asked about their secondary reasons, music and fun remained amongst the top three responses, followed by a broad idea of diversity that, in the qualitative responses, covered diversity in music, culture, sexuality, and performance or spectacle. When those surveyed were asked to elaborate further on why they personally enjoy watching Eurovision, only one respondent made direct reference to their British heritage and another participant of Italian background used this space to express their desire for an Italian victory; these were important points of expansion and clarification for these individual narratives but they did not constitute a significant trend in the survey data.

"Zombie myth" has emerged as an historiographical term to refer to dead ideas that persist despite evidence to the contrary (Stocking 2010), and the "European connection" narrative, as I have termed it here, emerges as the zombie myth of Eurovision in Australia. By using European migration as the constant reference point, we perpetuate that image of a white Australia that, as Gwenda Tavan (2005) has argued, experienced a long, slow death throughout the 1950s and 1960s, and we overlook the nuances of Australia as a multicultural society in the twenty-first century. Nevertheless, the link between immigration and Eurovision fandom ought not, admittedly, be declared entirely dead. Rather, we need to move beyond a myth fixated on a direct causal link toward a more nuanced understanding of how this immigration and its various attendant multicultural policies profoundly changed the way that Australians experience themselves and the world in which they live.

Migration is a feature of Australian society and remains the primary means of population growth. The ethnic composition of this migration flow has shifted over time, responding to push factors from abroad as well as Australia's internal pull factors. Some of these characteristics support the European connection narrative but others do not. Importantly, the narrative fails to account for the large portion of the Australian audience who do not lay claim to migrant or ethnic identities, or those whose heritage is not European. Indeed, the narrative presents a Eurocentric idea of multiculturalism (Gunew 2004) that risks perpetuating a "white nation" fantasy (Hage 1998). Preliminary findings from the Australian Cultural Fields project suggest that "ethnicity is not a singular determining factor in practices of cultural consumption" and that a more nuanced approach is required to understand patterns of consumption and taste in multicultural Australia (Noble and Ang 2018, 304). Similarly, my findings indicate that European ethnicities specifically are not a determining factor in Eurovision fandom in Australia. Policies of migration and multiculturalism do influence its presence and consumption, but the European connection narrative does not allow for a nuanced understanding of Eurovision's diverse Australian audience. Rather, Eurovision is a useful example for exploring the cosmopolitanising of Australian tastes as a result of the cultural diversity they encountered and negotiated in their everyday lives.

Many scholars of cosmopolitanism, including Stuart Hall (2003) and Jürgen Habermas (2001), suggest that the rise of multicultural societies create conditions in which cosmopolitanism can flourish, with the latter identifying a specific role for the media within this. As discussed in Chapter 2, this is the role that SBS television has played in the Australian media landscape, particularly as it developed more nuanced brand more focused upon ideals of transnationalism and cosmopolitanism than the delivery of multicultural media policy for migrant audiences. The presence of cultural and linguistic diversity on Australian television screens via SBS actively and successfully fostered a cosmopolitan habitus (Hawkins and Ang 2007); that is, it created a socio-cultural system in which Australians became more accustomed to difference and, as a result, more interested in it. The idea of a cosmopolitan habitus is also useful because it allows for a connection between the local and the global; diversity is something Australians value as part of themselves that they recognise, value, and seek out in their engagement with globalised culture. As one participant explains, "We're a nation of proud multicultural dags. And isn't that what Eurovision's all about?"

Almost one half of Australians are either born overseas or have at least one parent born overseas (ABS 2017b). The make-up of this migrant and second generation population is diverse; while the United Kingdom remains one of the top countries of birth according to the 2016 Census, it is followed by New Zealand, China, India, the Philippines, Vietnam, Italy, South Africa, Malaysia, and Germany, with most other national and ethnic groups from around the world finding some representation in Australia's migrant population (ABS 2017a). At the time that Eurovision came to Australia, the top ten birthplaces recorded in the 1981 census were entirely European (Cameron 1981), but the composition of the intake was beginning to change. European migration was certainly still dominant amongst regional origins, comprising 15.3% of the total Australian population, but had dropped by one percentile since the previous census in 1976; this one percentile change in the total population translates into an over ten percentile drop in the migrant population, from 81.3% European to 70.1%. The shift in migrant composition results from the complete abolition of racial restrictions from migrant selection in the 1970s and shifts in global politics and economics that altered external push factors for prospective migrants. Between the 1976 and 1981 censuses, Asian migration—also encompassing Middle Eastern nations in the census definition used at the time—had increased from 1.8 to 2.5%; this translates into a shift from comprising 8.8% of the migrant population to 11.6%. By the 1986 census, two Asian migrant groups, India and Vietnam, were amongst the top ten countries of origin (Castles 1991), and by 1996 Asian migrant groups represented 22% of all Australians born overseas (ABS 1997). These demographics paint a picture of a diverse Australia, but also a changing diversity. European heritages certainly dominated for much of the twentieth century, influenced by racially exclusionary immigration policy that in part comprised the so-called "White Australia policy," but by the time the Eurovision Song Contest made its way on to Australian screens—and by the time it was gaining traction amongst Australian viewers—this "European" idea of Australia had already begun to evolve into something even more ethnically, racially, and linguistically diverse.

It is certainly indisputable that the Eurovision Song Contest found its way to Australian screens as a result of the influence of European migration and the multicultural policy that responded to the diversification of Australian society as a result of mass migration in the wake of World War II—after all, SBS is the product of state multiculturalism. However, the large number of Australian Eurovision fans who can claim migrant heritage is a demographic reality rather than a causal link; that is, they are diverse because Australia as a whole is diverse, not because Eurovision is a text for migrant consumption. Furthermore, when we talk about "diverse Australia" or "ethnic diversity", we do not refer solely to European heritage. Many Eurovision fans who fall into this category of "ethnically diverse" are not necessarily from European backgrounds. Over one-fifth of my respondents claimed a variety of Asian heritages, and Australian diversity on the Eurovision stage has been almost exclusively represented by Asian and Indigenous Australians. One Filipino Australian participant first began watching the contest in 1988 when he and his parents encountered it by chance while channel surfing: "Coming from the Philippines, my parents and I were so used to (and missed) watching music and/or variety shows as they are constantly on Filipino networks. We stumbled across [Eurovision] on a Sunday night and I was immediately hooked." The experience of migration to Australia is part of this story but the participant and his family had no direct cultural connection to the contest nor the countries represented within it. Nevertheless, its format was familiar, and it was by migrating to Australia they came to know the contest. It is also important to remember that there are a large number of fans who do not claim any significant or immediate migrant heritage but who were drawn to the contest, like the Filipino family above, because of its particular brand of entertainment. Eurovision was (and is) fun, campy, and musical, and offered viewers a taste of diverse cultures and languages. For migrant and non-migrant fans alike, Eurovision offered something different from the regular fare on Australian television, something that may also evoke memories from or desires for a variety of elsewheres. Eurovision's presence in Australia emerges from historical trends in migration and resulting multicultural policy, but its appeal is evidently much broader.

Eurovision's peak in Australian popularity, as measured by television ratings, occurs in the 2010s but was building throughout the 1990s and early 2000s. Throughout this period, multiculturalism was evolving from a pragmatic policy addressing the special needs of migrants to a core element of Australian national identity to be celebrated by all, albeit not unproblematically (Jupp 2002; Hage 1998). Initially, multicultural policy of the 1970s and 1980s was a response to the ethnic diversity that emerged as a result of post-WWII migration to Australia. During that era, Australian immigration policy favoured white migration for the purpose of building a racially homogenous nation, but the influx of migrants from a variety of national backgrounds resulted in the proliferation of ethnic and cultural diversity, even if they were ostensibly white. The racial restrictions were slowly dismantled in the 1950s and 1960s (Tavan 2005), and were formally abolished in the early 1970s, although many scholars argue that much work remains to be done on completely banishing its legacy from Australian culture and institutions. Multiculturalism emerged as a settlement policy responding to the specific needs of migrants, many of which were being addressed in an ad hoc manner by grassroots movements and community organisations. The Galbally Report (1978) famously identified key areas of policy and service development, which included ethnic radio and television. SBS was developed from local Melbourne and Sydney radio stations into a national radio and television network designed to fulfil practical migrant needs, such as language preservation, English language acquisition, and circulation of information and news about Australia and homelands abroad, as well as the less tangible need of media representation. As Andrew Jakubowicz (1987, 18) observes, "the idea of a Special Broadcasting Service provided an opportunity to respond to the demand for cultural recognition that the ethnic middle class had for so long been denied in Australia." The desire for cultural representation in media was not, however, solved by the introduction of SBS. Lack of screen diversity remains a key area of criticism about Australian media (Screen Australia 2016), so the diversity offered by the artists selected to represent Australia emerges as one of the most valued dimensions Eurovision amongst my participants.

Multiculturalism as both a policy and a cultural ideal has been subject to much debate over its forty-plus years. Governments have variously claimed it as an integral part of Australian identity and distanced themselves from it as a problematic policy that prioritised special rights of migrants over "ordinary Australians." Despite observing a "discursive landscape of a 'troubled multiculturalism'" in official policy and various public and academic responses, Maree Pardy and Julian C. H. Lee (2011) have found multiculturalism to enjoy a fairly robust existence in the lived experience of Australians. In their research with migrant communities and organisations, multiculturalism was shown to be a "social fact" (Pardy and Lee 2011, 300) of most Australians' everyday lives. Drawing upon theories of cosmopolitanism and conviviality (Gilroy 2004), everyday multiculturalism emerged in the 2000s as a scholarly and empirical response to troubled multiculturalism. Importantly, everyday multiculturalism recognises multiculturalism as something applicable to all members of a culturally diverse society rather than simply as a minority rights policy. Yet it does not ignore the role of policy nor experiences of conflict, racism, and discrimination. Rather, everyday multiculturalism seeks to "understand how these wider structures and discourses filter through to the realm of everyday practice, exchange and meaning making, and vice versa" (Wise and Velayutham 2009, 3).

Multiculturalism is thus a social fact of contemporary Australian life. Elements of this, such as immigration levels, integration, and racism, remain important issues for public debate, but Australians have nevertheless grown accustomed to and accepting of cultural and racial diversity as part of their everyday social milieu (Markus 2017). SBS was instrumental in developing a cosmopolitan habitus amongst its Australian viewers, which in turn has affected perceptions of Australian society more broadly; that is, SBS and its programming contributed to helping Australians to understand and accept their diversity, and to foster a sense of connection to the rest of the world through this (Ang et al. 2008, 19–21). This perspective is, of course, subject to criticism, particularly as cosmopolitanism is often associated with the elitist consumption of difference (Hannerz 1992; Ang 2001), but cosmopolitan theorists since the 1980s and 1990s have sought to dislodge its association with elitism by placing its dynamics and ethos into more local and vernacular terms (Werbner 2006a, b, 2011); these vernacular theories inform both everyday multiculturalism and the concept of the cosmopolitan habitus.

SBS itself shifted toward a more "cosmopolitan aesthetic" in the 1990s (McClean 2014). This involved an accentuation of its "cult" programming, such as arthouse films, and quirky or kitsch television, such as the Eurovision Song Contest (Ang et al. 2008, 103–106); this was the aspect of SBS best associated with former Eurovision commentator Des Mangan. The shift toward this aesthetic was criticised for de-politicising difference and marginalising its working class ethnic base in favour of a new audience of urban cosmopolitan elites (McClean 2014, 51). While this is a legitimate concern, it presumes a particular subject position for migrants centred upon fixed authenticity and even victimhood. First, we must recall Jacubowicz's observation above that SBS was a response to the desire for recognition voices by the ethnic middle class. Second, the implied desire for an "authentic" ethnic subjectivity ignores the fluidity of cultures and socio-economics. Trends in intergenerational class mobility amongst migrant groups were already evident when SBS's cosmopolitan aesthetic was developed in the 1990s. Furthermore, the aesthetic was a recognition of a new market reality that emerged as a result of Australia's increasing diversity and the value placed upon diversity as a form of cultural capital: migrants and their children are now equally consumers of cosmopolitan cultural products available on a globalised market. Tastes hybridise as consumers—both migrant and longer term residents—access new products, and Eurovision is a good example of what Adrian Athique (2016, 106) terms a "crossover product," wherein a text deemed "ethnic" finds mainstream appeal.

Many fans adhere to the narrative that Eurovision's Australian popularity as a direct result of migration and multiculturalism, but they do not see its appeal as exclusive to a migrant demographic: "I think for a lot of Australians it's still an 'immigrant thing'. Certainly the harder-core fans I know are more likely to say they're 'Greek-Australian' or 'Italian-Australian' or whatever. But it's also just undeniable fun." Through reference to how other fans he knows, this participant reveals his own circle as diverse and Eurovision as a broad church that can appeal to all of them. Importantly, fans identify with the diversity of Australian society regardless of their own background. Take, for example, one respondent's explanation of Eurovision's appeal: "Lots of us come from European backgrounds." Her use of "us" here is telling; her demographic data reveals no claim to a migrant heritage, but in her response she uses inclusive language that presents an understanding of the collective Australian populace as ethnically diverse. Proximity emerges as something as important as personal heritage: "Even if you're not from a European background, you know someone who is or will have an affection for one of the countries." Interviewees, like the survey participants, all included multiculturalism as part of their rationale for Eurovision's popularity in Australia, but only Ana, Johan, Julie, Harry, and, to some extent, Tracy articulated this in varying ways as part of their personal narrative of fandom. Jordan, Drew, and Ryan noted migrant heritage from Europe and Oceania but did not express overt identification with these ethnicities, nor did they connect it in any meaningful way to their love of Eurovision.

Johan and Julie were the only two interviewees who had been fans of the contest prior to migrating to Australia from Sweden and the UK respectively. As such, Eurovision fandom was part of their pre-migration identity and they often do still consider it through their original national register, but not necessarily with nostalgic longing. For Julie, Eurovision has been a constant in an adulthood spent living in a variety of different places but she emphasises, "Eurovision doesn't make me yearn for the UK, or whatever, because a lot of the time the UK songs are not my favourite anyway." She describes any support she gives to the UK as "childish nationalism," suggesting both that nationalism itself is a form of childishness and that nationalistic attachment to English is simply an artefact of her own childhood in England rather than a real feeling she holds as an adult. Instead of making her yearn for home, something Julie claims to never have really felt anyway, Eurovision has continued to be a prompt for her wanderlust and her interest in other cultures and languages: "As I say, it doesn't bring out a sense of home or yearning for home or whatever else…it was often more along the lines of, oh, Stockholm looks cool, Tel Aviv looks like a great place…rather than thinking that shows home."

Similarly, Ana enjoys Eurovision for the way it makes other cultures and languages accessible, including her own Greek culture. In many ways Ana embodies what many other Australians would consider to be a typical Australian Eurovision fan. She is Greek Australian, a 1.5 generation migrant—that is, someone who was born overseas but migrated to Australia during early childhood (in Ana's case, infancy)—and Eurovision has become a key tool in maintaining an ongoing sense of attachment to her place of birth. She displays a strong sense of her Greek heritage and was particularly proud that the 2016 Greek entry was sung in her family's Pontian dialect. Yet even for Ana, who as a university student of language and history is very conscious of issues of culture and ethnicity, her primary reason for watching Eurovision is that "it's basically a fun competition." Her ethnicity informed much of our conversation but was offered—almost in an off-hand manner—as a secondary reason for her fandom: "Other than that [the fun competition], I guess being Greek, it's always been a nice thing to see your culture or your background up there with everyone else." Later in the conversation, Ana further emphasised the almost coincidental nature of her ethnicity and her Eurovision fandom: "Even though I'm Greek and also I'm Australian, that's not the big thing for me. The big thing is being part of this competition that brings together all of Europe. It's very cool. And the thing is, my favourite songs aren't Greece's song, or aren't Australian songs. They're random ones. One year it might be Azerbaijan, one year it might be Malta. It doesn't matter." Ana clarified that she experiences pleasure when Greece or Australia perform well, but that she doesn't support them purely on the principle of ethnicity or nationalism, as she feels other fans of migrant heritage may do. For Ana, a cosmopolitan ethos is more important than a specifically ethnic or national belonging.

Other fans with migrant backgrounds expressed similar sentiments: they were interested in seeing what their heritage country submitted to Eurovision, but their ethno-nationalism was usually tempered by other factors more important to them, such as Julie's curiosity about other languages and cultures, Ana's cosmopolitan idealism, or Jordan's taste for good pop music. In our conversation, Jordan approached his ethnicity as a demographic rather than cultural reality, as did Drew. Jordan's Italian background has played little to no influence on his Eurovision fandom nor how he assesses entries in the contest: "I judge entries by the standard of the song, not who sends the song." Furthermore, it had no bearing on his initial or ongoing attraction to Eurovision. It came from his own musical taste rather than an inherited or family tradition; he first started watching in the early 2000s as he was a fan of t.A.T.u. and introduced the contest to his Italian mother in the process. Similarly, Drew's Maltese Australian family didn't start watching Eurovision until he became a fan through exposure at a friend's dinner party: "My parents were babies when they came to Australia so grew up quite

Australian. It wasn't till I got exposed to Eurovision and I started pumping them up and getting them to watch it. We were quite a traditional Australian family with a staple of football grand finals, ARIAs and *Hey Hey It's Saturday*." Drew does, however, acknowledge that Eurovision is an opportunity to explore this part of his family background and has become more interested in Malta as his fandom in the competition has developed, but emphasises, "I was born here [in Australia], I've never really thought of myself as bring patriotic to anywhere else." For Drew, Eurovision has become an important way of connecting to the rest of his family and one day may become an opportunity for exploring some of his cultural heritage, but his foremost attachment is to the event as a form of musical exposure and entertainment that can be shared socially with friends and family.

Many fans also develop affective attachments to specific countries that they do not share heritage with, often through travel or education experiences, or even through exposure to Eurovision itself. Fans will watch for these countries above others and may be slightly more invested in their performance at the contest. For example, Daniel, who jokingly apologised multiple times during his interview for being a straight male WASP, expressed a soft spot for Belgium because he felt they placed a high value on culture, while Tracy and Harry became invested a little more in the Eurovision fortunes of France and Estonia respectively through marriage to people from those countries.

Initially, Harry was drawn to Eurovision by his own German background, but his ethnic attachment to Germany was not so great that he was not able to "put them over to one side" when he felt they began "entering some complete dreck." Harry's Estonian wife is also a Eurovision fan, but he differentiates between their levels of fandom as akin to "the difference between being the obsessive sports fan and the casual sports fan," as she doesn't devote as much energy as her husband does to following the contest outside of May each year. While Harry is certainly invested Estonia's performance, this seems slightly more crucial to his wife's enjoyment and, unlike Harry, she is unable to abandon her ethno-national association when dissatisfied with the quality of the songs; in fact, disappointment in the quality seems to heighten it. To illustrate his meaning, Harry described his wife's mortification at hearing Estonia's 2008 entry, "Leto svet", on public transport during one of their visits:

> I loved [the song] because it was in a whole bunch of different languages. A lot of it was in Serbian, and I'd been in Serbia, I'd been in Croatia, and a lot of it sounded like the way I was speaking whatever you want to call the language they speak in Serbia, Croatia and Bosnia. So I could see the humour in it, so I was sitting in the bus singing, while [my wife] was sitting there going, 'We'll never speak of this song again, it was terrible, it was embarrassing, it was awful,' and so on. Which it was.

The differences between their fandoms are more complicated and nuanced than Harry's "obsessive" and "casual" descriptors may suggest; while Harry may be more interested in Eurovision as a whole, his wife's ethno-national experience of the contest is far more acute. Despite any attachments that Harry has developed, his approach to the music is far more objective in that the entries do not reflect upon his national identity. Furthermore, Harry was also able to set aside his own heritage on the ground of poor quality. Conversely, his wife's sense of national pride was more tightly bound to the quality of Estonia's Eurovision submissions; she can enjoy the contest as a whole, but feels that Estonia's entries are a reflection of her national culture and identity (see Jordan 2014). Interestingly, Harry is not supportive of Australia's participation, describing it as "a novelty that has gone on too long," but also expresses clear relief at the quality of performances and the image of national identity submitted by Australia.

The multiculturalism deployed within Australian participation as a deliberate strategy in national representation will be discussed at greater length in the following chapter, but it is important to consider here how Australian fans engage with the concept of multiculturalism via the selected performers. This further illustrates the extent to which Australians have accepted and integrated multiculturalism into their sense of national identity and culture. All of the selected artists, from Jessica Mauboy's interval act to Isaiah Firebrace (and back to Mauboy again in 2018), are of migrant and/or Indigenous descent. In our discussions and in the survey, participants often used common discursive tactics of differentiating between "typical Aussies," meaning the blonde, surfy stereotype, and "multicultural Australia," but did not seem to suggest that one stereotype was any more or less worthy or authentic than the other. That said, most fans don't really expect their multicultural broadcaster to select an artist who embodies the Aussie stereotype in preference to an artist with a culturally diverse background. As Daniel observes: "You wouldn't have expected anything else from SBS….essentially, we've picked really good pop singers with what I will loosely refer to as a multicultural background—certainly not Anglo background." Only Julie expressed surprise at the ongoing visible diversity of contestants and voting ambassador Lee Lin Chin: "I know the SBS is sort of the diversity channel, so to speak, but [it was] actually quite unexpected, not a bad thing." One fan does, however, express concern that the artist selection has developed a new Australian stereotype:

> gradually as Dami Im, and later Isaiah was chosen, it has become a little disappointing that Australia has only been represented by people of Asian or aboriginal [sic] descent (including Jessica Mauboy in 2014). My European ESC friends jokingly ask every year which non-Australian artist (to them) Australia will select, and are starting to wonder whether Caucasian Australians exist, providing a skewed representation of real Australia.

This participant's relationship with multicultural Australia is complicated and suggests a problematic relationship between race and national identity. On the one hand, he suggests that Europeans hold an inaccurate idea of Australians that excludes those of non-white, but on the other hand he also suggests that the multicultural and postcolonial image being projected is not the "real Australia."

This view was, however, in the minority. Opinions of individual songs were often divided but the participants were generally satisfied with the "Australianness" of each individual artist. That is, they felt the artists' cultural diversity was an appropriate and accurate representation of contemporary Australia that they were happy to see displayed on the world stage. Fans praise the artists for being "good embodiments of what an Australian is" and for "show[ing] the diversity of Australia," frequently describing them as "ambassadors." Concerns focus more on the songs and the selection process than the artists and their suitability for representing Australia on an international stage. For example: "I would prefer contestants who did not come from tv talent contests. I am happy with the racial diversity of the contestants but I have found them to be boring and their songs bland and ballady with the exception of tonight again [sic] which was a nice boppy song." Although most participants praised Isaiah Firebrace's talent in general, his entry is identified by the fan quoted here and many others as the weakest offering to date. Despite their criticisms, the participants were also slightly defensive of him, because of his youth (Firebrace was seventeen at the time of his appearance) and his identification with a marginalised group in Australian society: "Isaiah was a good choice and it was great to see a young Indigenous voice represented, but he didn't quite pull off the song." In some ways, attitudes to Firebrace are similar to those towards Myf Warhurst as host: fans temper their criticisms with complements and good will because of other affective associations with the individual, as is the case with Warhurst, or the group that they are seen to represent, as can be seen in response to Firebrace.

The characterisation of contestants as ambassadors highlights the role Eurovision plays in fostering transnational connections between Australia and abroad. The effects of the cosmopolitan habitus are not limited to familiarising the foreign within the local, but in challenging parochialism more generally. Toby Miller's (2007, 18) observation of Australian media encapsulates this habitus: "Last I looked, Australia had a population of nineteen million people, but they seem to think it's worth knowing about billions of others." The desire to learn more about and to feel greater connection to other cultures through the Eurovision Song Contest is a strong theme in participant responses. As mentioned above, Julie credits Eurovision with fostering her interest in other languages and cultures that culminated in tertiary qualifications in linguistics, but the song contest was also one of the few ways that she could easily begin to know the world around her beyond the maps on her bedroom wall growing up in northern England:

> back in the 1970s and actually back in the 80s as well, if you wanted to find out what Jerusalem looked like or what Stockholm looked like, one of your only opportunities was the Eurovision Song Contest. Or if you ever wanted to hear Norwegian, unless you had family from Norway or your neighbour was Norwegian, which was very unlikely, you'd never hear it.

The internet and increased ease and affordability of global travel has changed this for many but it may still be the only way that many Australians come to think about, let alone know, Moldova or Azerbaijan; this is the practical corollary to the importance placed by nations on Eurovision as a form of national exposure (Jordan 2014, 2015a).

Although this might also promote awareness of political and economic situations in these countries, many Eurovision fans value the cultural exposure that may not be readily available elsewhere. Joanna explains:

> I also really appreciate that it kind of exposes me to music and culture beyond what I'd necessarily be exposed to otherwise. Because even though you know I could go digging on the internet to find out what's cool in Romania at the moment, you know, it's not necessarily something you're going to come across so Eurovision is a good way to discover really great artists.

This idea of cultural exposure aligns with Athique's (2016, 105) idea of "mediaculturalism…the visual equivalent of an overseas holiday or learning to enjoy foreign foods," and has long-lasting impacts on fans' consumption habits. Eurovision fans frequently purchase albums by former Eurovision artists, and algorithms used by music streaming platforms like Spotify facilitate even greater exposure to international musical cultures and trends.

While a sense of global belonging ranked seventh in survey participants' primary reason for loving Eurovision and fifth in their secondary reasons, their qualitative responses throughout the survey revealed a strong appreciation for the cosmopolitan dimensions offered by the Eurovision Song Contest. This spectrum ranges from an appreciation of others' cultures as another option in cultural consumption, as seen in Joanna's response, to a point of education or understanding, highlighted by Julie, to a more nuanced and meaningful sense of connection ("I feel connected to the rest of the world when I watch Eurovision") and belonging with the potential for social and political transformation ("I love the music, the sense of being a part of something worldwide that includes Australia, and the LGBTIQ community"). As demonstrated in the last two quotes, some responses suggest that Australians feel their nation and culture can often be excluded from global events and considerations, particularly when asked to elaborate upon the appeal of the contest to Australians more broadly. The feeling of connection during the contest suggests the absence of this outside of Eurovision.

Much of Australia's national psychology has been shaped by the so-called "tyranny of distance" from its ostensible ideological motherland (Blainey 1991). This sense of isolation, distance, and exclusion is best articulated in the following fan response:

> I personally feel that Australia is extremely isolated from the rest of the world. We wake up when everyone else goes to bed, and we have summer when they have winter and vice versa, but it's not just that, it's a lot of things that make me feel like, as Australians, we're…a bit excluded. So I'm extremely happy that Eurovision has invited us, that we get to share talented artists that were previously only known to us, that we get to show the world that we have stuff going on down here, we have talent and we're not just some hot country with dangerous animals.

Participation in Eurovision becomes a means of mitigating this sense of isolation and exclusion. Most importantly, this participant emphasises the idea the act of sharing fosters belonging and community not just with fellow Australians but with the rest of the world. Even Jessica Mauboy expressed this on the 2018 blue carpet: "This event makes you feel like you are really part of the world."

Eurovision's presence in Australia was facilitated by multicultural media policy designed to serve its diverse migrant population, but its ongoing appeal for Australian audiences is more indicative of a shifting sense of the multicultural national self than an ongoing desire for connection to migrants' European past. Both SBS and its audience evolved, and so too did the place of Eurovision within Australian society. The song contest becomes something Australians understand how to enjoy because of their lived experiences in a diverse society and a desired connection to the rest of the world that is not predicated upon their own cultural heritages. Understanding Eurovision and accepting non-white, non-Anglo-Celtic performers as representative of "us" is exemplifies the cosmopolitan habitus, in which Australian have been not just accustomed to but fluent in cultural diversity.

The case of Eurovision in Australia demonstrates the important role that communications technology and shared cultural texts can and should play in developing more meaningful transnational connections within this globalised world. Importantly, Australian Eurovision fandom illustrates how ideas of multiculturalism and transnationalism are in no way limited to migrants and their immediate descendants. Rather, it illustrates the positive cultural impact that the presence of migrants can have on broader society, particularly when supported by an official policy of multiculturalism that encourages a view of diversity as something that the society *is* rather than something it *has* (Hage 1998; Athique 2016).

Nevertheless, as will be shown in the next chapter, multicultural Australia has its cake and eats it too in this regard. While fans accept diversity as part of the Australian identity, they are also aware of the strategic utility of this representation. Chapter 6 will focus more on how this representation functions symbolically in understanding Australia's place in the world.

References

Ang, Ien. 2001. *On Not Speaking Chinese: Living Between Asia and the West*. London: Routledge.

Ang, Ien, Gay Hawkins, and Lamia Dabboussy. 2008. *The SBS Story: The Challenge of Diversity*. Kensington: UNSW Press.

Athique, Adrian. 2016. *Transnational Audiences: Media Reception on a Global Scale*. Cambridge: Polity.

Australian Bureau of Statistics (ABS). 1997. *4102.0—Australian Social Trends, 1997*. Canberra: Australian Bureau of Statistics.

Australian Bureau of Statistics (ABS). 2017a. *3412.0—Migration, Australia, 2015–16*. Canberra: Australian Bureau of Statistics.

Australian Bureau of Statistics (ABS). 2017b. Census Reveals a Fast Changing, Culturally Diverse Nation. June 27. http://www.abs.gov.au/ausstats/abs@.nsf/lookup/Media%20Release3.

Blainey, Geoffrey. 1991. *The Tyranny of Distance: How Distance Shaped Australia's History*. Rev. ed. Sydney: Pan Macmillan.

Cameron, R. J. 1981. *Census of Population and Housing, 30 June 1981: Summary of Characteristics of Persons and Dwellings*. Canberra: Australian Bureau of Statistics.

Castles, Ian. 1991. *Multicultural Australia*. Canberra: Australian Bureau of Statistics.

Galbally, Frank (Chair). 1978. *Review of Post Arrival Programs and Services: Migrant Services and Programs*. Canberra: Australian Government Publishing Service.

Gilroy, Paul. 2004. *After Empire: Melancholia or Convivial Culture?* London: Routledge.

Gunew, Sneja. 2004. *Haunted Nations: The Colonial Dimensions of Multiculturalisms*. London: Routledge.

Habermas, Jürgen. 2001. *The Postnational Constellation: Political Essays*. Cambridge: Polity.

Hage, Ghassan. 1998. *White Nation: Fantasies of White Supremacy in a Multicultural Society*. Sydney: Pluto.

Hall, Stuart. 2003. "Political Belonging in a World of Multiple Identities." In *Conceiving Cosmopolitanism: Theory, Context and Practice*, edited by Steven Vertovec and Robin Cohen, 25–31. Oxford: Oxford University Press.

Hannerz, Ulf. 1992. *Cultural Complexity: Studies in the Social Organization of Meaning*. New York: Columbia University Press.

Hawkins, Gay, and Ien Ang. 2007. "Inventing SBS: Televising the Foreign." *Australian Cultural History* 26: 1–14.

Jakubowicz, Andrew. 1987. "Days of Our Lives: Multiculturalism, Mainstreaming and 'Special' Broadcasting." *Media Information Australia* 45: 18–32.

Jordan, Paul. 2014. *The Modern Fairy Tale: Nation Branding, National Identity and the Eurovision Song Contest in Estonia*. Tartu: University of Tartu Press.

Jordan, Paul. 2015a. "From Ruslana to Gaitana: Performing 'Ukrainianness' in the Eurovision Song Contest." *Contemporary Southeastern Europe* 2 (1): 110–135.

Jordan, Paul. 2015b. "Australia to return to the Eurovision Song Contest!" *Eurovision.tv*, November 17. https://eurovision.tv/story/australia-to-return-to-the-eurovision-song-contest.

Jupp, James. 2002. *From White Australia to Woomera: The Story of Australian Immigration*. Cambridge: Cambridge University Press.

Markus, Andrew. 2017. *Mapping Social Cohesion: The Scanlon Foundation Surveys 2017*. Caulfield East: Monash University.

McClean, Georgie. 2014. "National Communication and Diversity: The Story of SBS." In *'For Those Who've Come Across the Seas…': Australian Multicultural Theory, Policy and Practice*, edited by Andrew Jakubowicz and Christina Ho, 45–56. London: Anthem Press.

Miller, Toby. 2007. *Cultural Citizenship: Cosmopolitanism, Consumerism, and Television in a Neoliberal Age*. Philadelphia: Temple University Press.

Noble, Greg, and Ien Ang. 2018. "Ethnicity and Cultural Consumption in Australia." *Continuum: Journal of Media & Cultural Studies* 32 (3): 296–307.

Pardy, Maree, and Julian C. H. Lee. 2011. "Using Buzzwords of Belonging: Everyday Multiculturalism and Social Capital in Australia." *Journal of Australian Studies* 35 (3): 297–316.

Screen Australia. 2016. *Seeing Ourselves Reflections on Diversity in Australian TV Drama*. Sydney: Screen Australia. https://www.screenaustralia.gov.au/fact-finders/reports-and-key-issues/reports-and-discussion-papers/seeing-ourselves.

Stockings, Craig. 2010. "Introduction: The Walking 'Undead' and Australian Military History." In *Zombie Myths of Australian Military History*, edited by Craig Stockings, 1–9. Sydney: University of New South Wales Press.

Tavan, Gwenda. 2005. *The Long, Slow Death of White Australia*. Melbourne: Scribe Publications.
Werbner, Pnina. 2006a. "Understanding Vernacular Cosmopolitanism." *Anthropology News* 47 (5): 7–11.
Werbner, Pnina. 2006b. "Vernacular Cosmopolitanism." *Theory, Culture & Society* 23 (2–3): 496–497.
Werbner, Pnina. 2011. "Paradoxes of Postcolonial Vernacular Cosmopolitanism in South Asia and the Diaspora." In *The Ashgate Research Companion to Cosmopolitanism*, 107–123. Farnham: Ashgate.
Wise, Amanda, and Selvaraj Velayutham. 2009. "Introduction: Multiculturalism and Everyday Life." In *Everyday Multiculturalism*, edited by Amanda Wise and Selvaraj Velayutham, 1–17. Basingstoke: Palgrave Macmillan.

CHAPTER 6

From Mr Eurovision to Australian Idols: Australian Performances (and Performing Australia)

Abstract Australia's participation in the Eurovision Song Contest is hotly contested by media commentators and fans alike, but many agree that the Australian delegation has made an effort to submit quality songs and to present a consistent image of Australia as diverse. This chapter examines the various representations of Australia that have appeared on the Eurovision stage with a consideration of how this adheres to both national and international ideas about multiculturalism and diversity. It argues that the deliberate selection of artists with various Asian and Indigenous heritages disrupts the European connection narrative. By highlighting its postcolonial status and its more recent waves of migration, representations of Australia at the Eurovision Song Contest present an image of contemporary Australia poised to be a bridge between Europe and Asia.

Keywords Multiculturalism · Representation · Performance · National identity

In 2015 Australia was invited to compete in the Eurovision Song Contest for the first time in the contest's history. Guy Sebastian's wild card performance was intended as a one-off invitation to celebrate the Eurovision Song Contest's sixtieth anniversary, and to acknowledge thirty years of Australian fandom. However, towards the end of 2015 it was announced

J. Carniel, *Understanding the Eurovision Song Contest in Multicultural Australia*,
https://doi.org/10.1007/978-3-030-02315-7_6

that Australia would return in 2016, much to the displeasure of some. Sebastian's Eurovision entry had been preceded the previous year by a guest performance from Jessica Mauboy during the voting intermission, and in 2016 was succeeded by Dami Im, and Isaiah Firebrace in 2017. Mauboy returned to Eurovision again in 2018, this time as a contestant. Australia's Eurovision history and the presence of Australia and Australians at the song contest predates not only the appearances of Sebastian, Mauboy, Im, and Firebrace but also the beginning of the SBS telecast in the mid-1980s, but Eurovision 2015 was significant because it was the first time a performer represented and competed on behalf of the Australian nation. The initial and ongoing selection of the performers is therefore not just about showcasing Australian talent, but about defining Australian national identity on a global, but predominantly European stage. Significantly, as Australians who can each lay claim to a combination of various Indigenous, Asian and European ethnic heritages, Mauboy, Sebastian, Im, and Firebrace are visible minorities within Australian society, yet by representing Australia on the Eurovision stage they come to signify the multicultural Australian whole.

The inclusion of Australia as a competing nation in Eurovision is both novel and controversial, largely due to its obvious geographical distance from Europe. Australia is not, however, the only ostensibly "non-European" nation to participate in the contest; many have puzzled for years over, for example, the participation of Israel and Azerbaijan, nations that challenge popular understandings of what constitutes "Europe." Yet their geography is more proximate, if not included in definitions of Europeanness, and both nations have been tied more closely and materially with Europe in their contemporary politics and economics than Australia has been. Azerbaijan, for example, has been a member of the Council of Europe since the early 2000s and in recent years has sought to further strengthen its ties to various European powers, including the European Union. As Etienne Balibar (2009, 6) observes, the borders of Europe are "arbitrary" and the "character of the European 'continent'" is "unlimited," enabling it to encompass parts of the Middle East, such as Israel, and the Caucasus, such as Azerbaijan and Turkey. In his history of the idea of Europe, Anthony Pagden's (2002, 35) wry summary of key moments in the historical construction of Europe emphasises the interplay between west, east, and middle east: "an abducted Asian woman gave Europe her name; a vagrant Asian exile gave Europe its political and finally its cultural identity; and an Asian prophet gave Europe its religion." Yet even Pagden's inclusion of Asianness in ideas of Europe has continental limitations that excludes a separate continent in the Pacific.

Australia's inclusion may defy any tenuous geographical logic, but the reality of Eurovision participation is rather prosaic: any country that has a public broadcaster with full membership in the EBU are immediately eligible to participate in the contest, while associate members may apply to participate, subject to board approval. SBS has been a long-standing associate member of the EBU as its Eurovision network has been an important source of international news and sports. Other associate members include broadcasters in the United States, Canada, China, Japan, and Brazil, amongst others scattered around Asia, Africa, and the Americas. Australia is, however, the only geographically distant associate member to be invited to participate. Australia's first request to participate in 1970 was denied (Vuletic 2018, 49). The decision is understandable: at that time, Eurovision was an exclusively western European event, few Australians were aware of the event beyond a short line in the newspaper, and global media and culture had not yet involved into the complex network that it is today. Australia's success in joining the contest in 2015 was the result of several years of lobbying by SBS and its production partner BlinkTV, with Jessica Mauboy's interval act providing a test case for Australian inclusion. After four years in the contest, Australia is still not considered a full member of the competition; each year the question of its inclusion is considered by the EBU Reference Group.

Many accepted the appearance of Mauboy and Sebastian as gimmicks but were less sanguine when the invitation was extended again in subsequent years. BBC commentator Graham Norton famously opined that he found Australia's inclusion after the purported one-off of 2015 "kind of stupid" because it is "on the other side of the world." Norton acknowledged the relatively elastic boundaries of "Europe" taken by the contest and the EBU, but an expansion into the southern hemisphere was too much of stretch for him and others. Indeed, much media and fan commentary emphasised the "Euro" in Eurovision, although others pointed out the practical limitations of the expanding boundaries as the increasing number of competitors are making the contest too big to manage.

Geography aside, other critics expressed concern about how Australia's presence would affect the authenticity of the song contest. In a somewhat backhanded expression of nationalism, Monica Tan (2016), an Australian cultural commentator for the *Guardian*, suggested that the quality of the performances and costuming of the Australian entries—that is, that they are too good and too well-dressed—not only signals how little Australia belongs, but that they are an instrument of destruction for "Europe's zaniest singing competition." According to Tan, "Australia's prolonged participation is only diluting the crassness and bizarreness that makes Eurovision so delicious to watch." One participant aligned with Tan's view: "Eurovision works because of its quaint Euroness, so adding Australia and potentially others (like USA and China), you destroy that charm. Most of all, Australia doesn't add anything other than one extra song." Other participants disagreed, arguing that the illogic of Australia's participation is only further evidence of the contest's reputation for "craziness." One stated, "It's a fun new angle, causes some controversy, and gives a nod to the weird level of random here." Another pointed out that "clearly we haven't 'ruined' the competition, it's just as popular as ever." (This is not entirely true; both Australian and global audience numbers peaked in 2015 and have since declined slightly.)

Australian fans may express some reservations, but they are predominantly excited and supportive of Australia's ongoing participation. Over 60% of those surveyed reported moderate to extreme excitement (ratings 7–10 on a Likert scale) about Australia's participation in 2015, rising to 70% in 2016, and dropping slightly to 65% in 2017. Regarding Australia's continued participation beyond 2017, 69% of respondents agreed that Australia ought to participate in the future, with the remaining respondents split fairly evenly between disagreement and uncertainty. In explaining their responses, many participants did note that their excitement about confirmed participation has waned over the years. As one observed wryly, "'Australia's in Eurovision' has the same level of interest as 'Finland's in Eurovision'"; it has, in the words of another participant, become "a bit normal." It is difficult to draw any specific correlation between shifts in enthusiasm for involvement and the performers, as those surveyed report greater excitement for Guy Sebastian than for Isaiah Firebrace but record marginally more excitement about Australian participation in 2017 than in 2015. Australian participation is confirmed before the artist is selected and announced and in 2016 Australians were buoyed by Dami Im's extraordinary result, so may have been optimistic about future prospects in the contest. As with the commentary team, levels of familiarity may account for the public response to Sebastian, who has been part of Australian pop music and television culture since the early 2000s, in comparison relative newcomer Firebrace.

Excitement about involvement and support for Dami Im were relatively consistent in survey results. Interestingly, many fans reported that they were not initially convinced of the potential for success of "Sound of Silence" in the lead-up to the competition, but were won over by Im's powerhouse performances in the semi and grand finals: "'Sound of Silence' sounded rather average in the studio, but the live performance was incredible." Certainly, those I interviewed prior to the 2016 contest were overwhelmingly sceptical of its chances, dismissing the song as "another boring ballad." In follow up interviews after the 2016 grand final, many had shifted perspective on the song. Drew, for example, initially described the song as "boring…really disappointing" but after the event stated, "What a turnaround. From hearing the recorded version and being bored to seeing her belt it out live and the hairs standing up on my neck. She did us proud." Quality of song is important, as many fans see this as proof that Australia takes the contest "seriously," which is in turn justification for ongoing participation. As one participant summarises, "We have proven that we can take it seriously, provide high quality acts and big name artists, and have had good results."

Australian performers have been no strangers to the Eurovision stage. According to Chris West (2017), an Australian, Shirley Abicair, was slated to perform at the very first Eurovision in 1956 as the British representative, but Gordon Roxburgh and Paul Jordan (2017) have dismissed such stories around the UK's absence in 1956 as fan theories. Nevertheless, it is interesting that an Australian should be at all tied to this Eurovision mythology from the outset as it appeals to a sentimental desire for authentic, historical belonging to the contest. The belonging is tenuous at best, but an examination of past Eurovision performers, including the myth of Abicair, highlights the strong transnational links between Australia and the various Eurovision nations from which its large post-World War II migrant population derives. This in turn reveals the ongoing ideological and cultural connections that have been maintained between Australia and Europe. The history of Australians on the Eurovision stage reveals a microcosm of Australia as a migrant nation historically, and of Australians as mobile, global, and creative citizens. It highlights migration as a system of mobility that, particularly in an era marked by postmodern globalisation, is not one-way, and that is intimately tied up with cultural performance and exchange.

Johnny Logan, for example, is one of the most successful Eurovision winners. Having won a record three Eurovision Song Contests (twice as performer and once as songwriter), he is known as "Mr Eurovision." While Logan may not necessarily be categorised an Australian performer as such, having lived most of his life and forged his musical career predominantly in Ireland and Germany, he does constitute an important Australian connection with Eurovision. Such connections were publicised in media commentary during the lead-up to Guy Sebastian's appearance in 2015 (see, for example, Adams 2015), perhaps to illustrate how the inclusion of Australia in the contest was a natural progression of its ongoing, indirect participation. Logan's most famous win is 1987's "Hold Me Now." Born in Frankston, Victoria, Logan's birth name is Seán Patrick Michael Sherrard. (One fan suggests that the local Frankston council ought to erect a monument to Logan in order to "establish itself as the focal location of Eurovision in Australia.") He is the son of Patrick O'Hagan, an Irish tenor who found moderate success in Australia, including a short-lived 15-minute television slot called "Patrick O'Hagan Sings." When Logan was three his family returned to Ireland and, although Australian-born, he is considered a categorically Irish singer. The case of Johnny Logan exemplifies the particularly Australian habit of claiming Australians amongst the famous, yet it is not entirely misguided as his official website biography tells a story of strong transnational connection with Australia even after his family's return migration. Logan's father later toured Australia and New Zealand with a young Sean in tow, and the family later returned to Australia in the 1980s, where they lived at the Gold Coast, home to Twin Towns, where Logan has frequently performed.

While he is perhaps the most famous, Logan was not the first. He is preceded in his Eurovision infamy by one who was more recognisably Australian at the height of her career, Olivia Newton-John. Representing the United Kingdom in 1974, she came fourth to ABBA's "Waterloo." Newton-John migrated from the UK to Australia with her family in 1954, when she was five. Her success in Australia led to an opportunity to return to Britain, where she appeared regularly on the show of another Eurovision stalwart, Cliff Richard, and was eventually selected to represent the UK in the contest. Despite her British origins, Newton-John's Australianness was a core part of her brand in the 1970s; the character of Sandy in *Grease* was re-written to be Australian in order to accommodate Newton-John (News.com.au 2016). As such, Newton-John's performance in Eurovision in 1974 was, unlike Logan, recognisably Australian.

Other Australian performers in Eurovision have been less iconic in their homeland, although Gina G in 1996 experienced some success with her entry in the pop/dance charts of Australia, Canada, the US, and the UK. Jane Comerford, who represented Germany as the lead singer for Texas Lightning in 2006, is less well-known in Australia, having predominantly carved out her musical career in Germany's country music scene. Even in 2015, Guy Sebastian shared the stage with other Australian performers: Mary-Jean O'Doherty Basmadjan performed as part of Armenia's diasporic entry, Genealogy, and Katrina Noorbergen performed backing vocals for Russia's Polina Gagarina, having also co-written the song. Noorbergen's Eurovision-related credentials are expansive, demonstrating the potential of the contest to foster creative connections. She also wrote a song for former German contestant Ann Sophie's post-Eurovision album, co-wrote 2010 German winner Lena Meyer-Landrut's fourth album, *Crystal Sky*, and was the support act on that album's tour. (The Australian connection can be pushed - albeit very tenuously—even further through Meyer-Landrut, as she performed former *Australian Idol* contestant Lisa Mitchell's "Neopolitan Dreams" as part of her bid for Eurovision in *Unser Star für Oslo*.) Once songwriters as well as performers begin to be included, Australians start to feature even more, having also written songs for countries such as Germany and Malta. The presence of Australian fans and performers at Eurovision demonstrates the mobility of contemporary Australians, and the addition of the audiences at home in Australia is further testimony to their enthusiastic participation in international cultural events and the global culture industry. It also demonstrates a dedication to the Eurovision Song Contest specifically, which raises further questions about what Eurovision might offer Australia and Australians culturally, politically, and ideologically.

Examining the recent performances by Australians at the Eurovision Song Contest reveals a strategic display of visible cultural diversity that articulates a modern, cosmopolitan and multicultural identity. This in turn also references legacies of colonisation that have been instrumental in establishing and maintaining the tie between Australia and Europe, while at the same time self-consciously disrupting it. Both Paul Clarke of BlinkTV and Michael Ebeid of SBS confirm that the ethnic diversity of the performers was paramount in the selection process, although Clarke also emphasises that more prosaic considerations, such as scheduling and appropriateness of the song/performer match, are also important. It is also well-known that the Australian Eurovision entry is supported by

Sony BMG, which limits artist selection to those on their books (Molloy 2016). As discussed in Chapter 5, fans are unsurprised by the strategic diversity deployed by the Australian producers and broadcasters, and generally agree that the display of diversity is an appropriate representation of their national multiculture on an international stage, particularly as it disrupts global stereotypes of white Australianness. The ethnic and racial diversity simultaneously encompasses Australia's professed self-image and offers Europe an aspirational model for "unity in diversity."

The display of diversity is not, however, unproblematic. First, it recalls Ghassan Hage's (1998, 160–161) criticisms of multiculturalism as something Australia has, and displays, rather than something it is. Such a distinction has been expanded upon critically and theoretically in the notion of everyday multiculturalism, which focuses of the lived practices of diversity and its intrinsic value for a diverse populace rather than its instrumental value for either the state or the market (Wise and Velayutham 2009; Werbner 2013). Fan responses to Eurovision contestants do indicate an understanding of both perspectives: multiculturalism is a part of their everyday experiences and identities, but they also understand the pragmatic benefits of cultivating this image internationally. Second, the display of diversity presents a sanitised version of Australia's complex history of migration and colonialism. On the Eurovision stage, Australia is presented as unproblematically diverse. The image of success that Sebastian, Im, Firebrace and Mauboy each project is important to provide diverse role models to young Australians, but it also elides experiences of racism and discrimination experienced by Indigenous and migrant Australians. Mauboy did, however, disrupt this with her 2018 entry "We Got Love." Although the song's message is overwhelmingly positive, Mauboy spoke openly in the media about how it was partially inspired by reports of a two-year-old girl being sexually assaulted in the Northern Territory town of Tennant Creek. For Australians, this case recalls the controversial Northern Territory Emergency Response (known as the "Intervention") implemented in 2007 in response to the *Little Children Are Sacred* report, which resulted from an inquiry into sexual abuse of children in Aboriginal communities. Evaluations of the Intervention have found it to be a failure (Gray 2015). Not only have its practical measures been unsuccessful, it disrupted Indigenous agency and further damaged historically fragile relationships between Indigenous Australians and the state (Scott et al. 2015). Mauboy is active on Indigenous issues in Australia, but to speak openly of these issues to global media is important as she problematises the international image of happy postcolonialism that her success may suggest.

It is significant that the performer who first takes to the Eurovision stage as an Australian representative (albeit non-competing) is Indigenous, as well as of a non-European (West Timorese) migrant background. Interestingly, as Celia Lam (2018, 120) has noted in her analysis of how Australian identity is constructed around Australian Eurovision entrants by the media, Mauboy's identity is always framed in terms of her indigeneity rather than her status as a second generation migrant Australian. Certainly, Mauboy is complicit in this construction, but it may also reflect her stronger personal identification. Dami Im's identity is frequently framed as being "Korean-born" and living in Brisbane rather than as Korean Australian. Lam (2018, 120) observes that mentioning Im's place of birth "in passing re-emphasises the primary lens through which Im is viewed; as a resident of Brisbane and Australia." Such framing works to acknowledge Im's racial otherness, yet stake a claim for her belonging to Australia. Conversely, Guy Sebastian is usually referred to simply as "Australian," without further qualification. Lam suggests that this could be due to the longevity of his celebrity or the complexity of his mixed heritage. Sebastian migrated to Australia from Malaysia at age seven, and his parents are of Malaysian-Sri Lankan Tamil and English-Portuguese descent. A notable exception to this comes from SBS. Writing on SBS's Eurovision page, Siobhan Hegarty (2015) describes Sebastian as a "migrant musician [who] is basically a walking, well-used passport." This description foregrounds Sebastian's migrant status, yet also frames him as experienced and mobile. Isaiah Firebrace is outside the temporal limits of Lam's study and his identity is arguably less complicated than the other Eurovision alumni; reviewing the media coverage of his Eurovision performance in 2017, he is generally described in terms that emphasise his Indigeneity and his origins in regional New South Wales.

That each of Australia's Eurovision contestants are alumni of national reality television music competitions is interesting for considering whether Australia is or simply has diversity. Guy Sebastian won the very first season of Australia's local version of the global *Idol* franchise in 2003 (now defunct), while Jessica Mauboy came second place in the fourth season in 2006. With the exception of dating programs such as *The Bachelor*, reality shows are frequently notable for the diversity of their contestants. As Osman Faruqi (2017) observes, "diversity on screen has been quarantined to shows that feature the same traits people of colour have been stereotyped as exhibiting for over 100 years: cooking and performing for white people." On the one hand, the implications of structural power are as problematic as Faruqi suggests. On the other hand, that diversity appears on commercial television, that it is framed in a positive light, and that it is positioned as part of Australian "reality" is invaluable. As Joanna

McIntyre (2017, 88) emphasises, Australian reality television is certainly no "utopia of affirmative action," but it can be useful for repositioning a variety of racial, cultural, and gender norms both on- and off-screen. *Australian Idol* was the first to achieve this. Of the twelve performers cast in its inaugural season, five were non-white, including Guy Sebastian. Furthermore, Sebastian was victorious against the more stereotypically "Aussie" Shannon Noll. Mauboy also appeared on *Idol*, losing to Irish migrant Damien Leith. Dami Im and Isaiah Firebrace are both *X-Factor* alumni, and Australia's first and second Junior Eurovision contestants, Bella Paige and Alexa Curtis, emerged from Australia's reality/talent circuit in Channel Nine's *The Voice Kids Australia* (as 2014 runner up and winner respectively), and are similarly representative of a diverse Australia due to their Macedonian (Paige) and New Zealand (Curtis) heritages. As reality television programs, these various music talent programs showcase cultural diversity as part of Australian contemporary reality, as a "social fact" (Pardy and Lee 2011, 300).

Cultural diversity has emerged as a de facto feature of these music talent shows. Even viewed with some healthy cynicism about casting and production in reality television, this is simply a result of what happens when you canvas talent from "ordinary Australians." It is also striking to see this diversity on commercial programming in Australia, as this is often criticised for its overwhelming lack thereof (Screen Australia 2016). Consequently, the presence of diversity on *Australian Idol* and similar programs is more disruptive to the status quo than that found on SBS because it appears on mainstream commercial television rather than a marginalised, government-subsidised network mandated to engage with diversity (Huijser 2007, 137). As a culturally diverse display of talent is not explicit in the goals or remit of these programs, the diversity of the casts becomes representative of the multiculturalism that Australia *is*, rather than it *has*. Due to their successes in the mainstream television and music industries of Australia, Sebastian and Mauboy both emerged as highly recognisable Australian figures and oft-touted symbols of its diversity, and as such were suitable inaugural representatives for Australia at Eurovision.

As with all Eurovision contestants, Mauboy and Sebastian shared important roles as cultural ambassadors for Australia but the different styles and tones of their performances work to articulate Australia's shifting place within the Eurovision Song Contest as it moved from guest to competitor. Mauboy's performance differs from that of Sebastian in that while Sebastian was *competing for* Australia as a wildcard entry, Mauboy was intended to be *symbolic of* Australia. Preceded by a deliberately campy introduction, Mauboy's "Sea of Flags" is openly nationalistic. By performing the Australian nation in this appearance, Mauboy establishes a national space for Guy Sebastian's competitive entry, freeing him up to represent the Australian culture industry sans overt nationalist display.

Indeed, the entire Australian contribution to the 2014 interval act is a self-conscious display of national image and identity that is both playful (in the camp introduction) and sincere (in Mauboy's performance of "Sea of Flags"). Mauboy's song is preceded by a pre-recorded introduction where ESC Executive Supervisor Jon Ola Sand and his fellow producers mock Australia's dream of performing at Eurovision (Eurovision Song Contest 2014). Standing in front of the Sydney Opera House, surrounded by Australian Eurovision fans, commentator Julia Zemiro begs the producers to allow Australia to perform. When they are told that Australia cannot participate because they are not part of Europe, Zemiro airlifts Australia over to the European continent. This is followed by a short song and dance piece featuring playful stereotypes of Australianness: surfers, singlet-wearing beer drinkers, koalas, workmen, Australian rules players, a kangaroo in a Socceroo outfit, a hammerhead in a ballet tutu, and several drag performers. The singers announce the arrival of the new country: Eurostralia. The animation nestles the Australian continent into the western coast of Europe, re-arranging its geography completely and notably ousting the UK. From a European perspective, Australia's participation is largely a novelty. From an Australian perspective, it is the culmination of a long-held dream—and not necessarily one that is centred literally on Eurovision; like other nations that exist on the geographical edges of Europe, such as Azerbaijan, Australia had the opportunity to assert its fantasy of European belonging. The Eurostralia performance is a camp display of popular stereotypes of Australianness, while the animation symbolically closes the geographical distance that has separated Australia from the European motherlands for the centuries since European colonisation. If we are also to read both Mauboy's and Sebastian's performances as "auditions" for continued involvement, Sebastian's performance needed to emphasise Australia as *member* rather than as *novelty*. Compared to the deliberate cultural references and camp prelude found in Australia's 2014 guest appearance, Sebastian's 2015 performance is, as Claudia degli Alessandrini (2015, 47) describes, "culturally-neutral" and "sober." This, Alessandrini argues, "shows that the intention was not to portray Australia as distinguished and different from Europe, but rather as homogenised or blended into the western pop-song tradition. The performance then serves to prove that Australia is not Europe's 'other' but almost an extension of Europe." Consequently, without the camp fanfare that preceded Mauboy's performance in 2014, Sebastian's "Tonight Again" (Eurovision Song Contest 2015) aligns with some of the more "serious" Eurovision contributions that aim to win the contest on broad musical appeal—for example, Loreen (Eurovision Song Contest 2012)—rather than novelty—Lordi (Eurovision Song Contest 2006). While camp entries are still part of the Eurovision tradition, these are now in the minority as there is an increasing trend toward using the contest as a competitive global outlet for quality music, and Australian entries can be understood not as submitting to homogenisation, as Alessandrini suggests, but as understanding the competition as a serious opportunity for developing new markets for Australian talent.

The lyrics of Mauboy's song, "Sea of Flags" (co-written with Ilan Kildron and John Stuart Crichton), deliberately evoke Australia's postcolonial multiculturalism, personified in its performer. Stephanie B. Guy (2015, 12) identifies Jessica Mauboy as one of several contemporary Aboriginal musicians who fracture Australia's constructed nationalism and the "non-Indigenous reliance on the noble savage as the essential symbol of Aboriginality." Young, stylish and affable, Mauboy represents a modern and cosmopolitan Indigenous identity. Mauboy has described the song as a "cultural statement" that sought to "create a landscape and an introduction, a view, of this country" (quoted in Idato 2018). The "sea of flags" of the title and lyrics refers simultaneously to the flags waved by the Eurovision audience below and to the many migrants who have helped to constitute Australia as a multicultural nation, while the lyric "dancing through an ancient dreaming" refers to Australia's Indigenous culture. This is further emphasised by Mauboy's stylised dance moves and gestures that reference traditional dance; this is perhaps now a personal signature and signalling of identification for Mauboy at Eurovision, as she also incorporated Indigenous-inspired dance into her performance of "We Got Love" in 2018. The performance demonstrates how, as Guy (2015, 4) argues, the problematic binary of "traditional" and "contemporary" applied to Indigenous music can exist simultaneously and symbiotically.

The last time that Australian indigeneity had been referenced in any way at the Eurovision Song Contest was in Austria's 1977 entry from Schmetterlinge, "Boom Boom Boomerang," with lyrics featuring rhymes that played upon Indigenous terminology, such as "boomerang," "kangaroo," and "didgeridoo." The song itself had nothing to do with Australia or Aboriginal culture, but is considered to be the first entry that did not take the contest seriously—in other words, the first novelty entry. The song was a deliberate poke at the artistic integrity of the contest and the music industry more broadly, mocking previous Eurovision entries. Although the powerhouse successes of Lena, Loreen and Conchita in recent years has, to some extent, revitalised the contest as a fairly legitimate platform for performers to launch European, if not global, careers, the contest has been derided for years as itself a novelty and a parody; this is one of the reasons Wogan cites as his reason for abandoning the commentary, and was implied also in 2017 winner Salvador Sobral's controversial comments about "disposable…fast food music." Viewed now as part of Eurovision's increasingly camp history, "Boom Boom Boomerang" lacks the good-natured irony that generally can be found in Eurovision performances. The choice of lyrics is dismissive of Australian Indigenous culture; they are chosen for their seemingly nonsensical qualities, making the song more like a novelty children's ditty by Rolf Harris. While the target of the mockery is the European music industry, the effect—even if unintentional—is the infantilisation of Indigenous culture (Ramm 1989; Edwards 2009, 33–34).

Australian delegations have taken themselves and the contest more seriously than Schmetterlinge. This counteracts elements of novelty in the Australian participation while also recognising the legitimacy of the European music industry. With the exception of Firebrace, whose fame was relatively new, Australia sends established and relatively successful performers in hope that they will find further success in the European market. Furthermore, as a relatively new competitor and one with more tenuous belonging, there is also more pressure to send higher quality acts. As head of the Australian delegation Paul Clarke (quoted in Van Ee 2017) says, "We are on probation, and if we don't come up with a good artist, maybe they will just sack us. We know we need to bring it, we need to be the Sweden of the south, to be able to stay in the contest. That is how it feels to us." Part of the reason for Australia's earnestness in approaching the competition rests in that same dynamic of marginalisation experienced by states that are geopolitically peripheral to Europe; these marginalised states take Eurovision far more seriously because they see it as a platform to perform the nation and to seek European belonging (see, for example, Sieg 2013; Ismayilov 2012). Where Schmetterlinge sought to deride the contest and its effects on the European music industry, Australia seeks to belong to the Eurovision community as a specific global cultural enterprise and to launch European careers for its performers. It is assumed within this that Eurovision is a legitimate and serious contribution to the very European music industry derided by Schmetterlinge. Furthermore, Mauboy, Sebastian, Im, and Firebrace appear at Eurovision, and its attendant media circus, not just as performers but as cultural ambassadors for Australia.

The method of artist selection is both an accident of Australian multiculturalism and migrant history and deliberately playing to the national Australian (and SBS network) "brand" that has evolved from this. Nation branding is, after all, a key function of how Eurovision performances are constructed, performed, and sold (Jordan 2014). Such nation branding can also be seen in the multicultural former commentary team of Pang and Zemiro, as well as the selection of the voting reporters, Lee Lin Chin, a Chinese-Indonesian Australian journalist and fashionista with a cult following, and Chin's successor, Portuguese Australian news reporter Ricardo Gonçalves. The racial identity of the various Australian representatives is also important. As Lutgard Mutsaers (2007, 62) has observed in reference to blackness and Dutch performances at the song contest, "Eurovision functions within a visual medium, and therefore looks are relevant in the overall picture of a national entry." Whiteness, she argues, "is still taken for granted." Any distruption to this whiteness is therefore a significant commentary on racial identity and Europeanness. Accordingly, the decision to foreground non-whiteness—and specifically Asian and Indigenous identities—is equally a commentary on Australian national and racial identity, and a deliberate, visual claim to diversity. Importantly, it is a disruption to the European connection narrative that privileges ideas of white Australianness and white multiculturalism.

Australian multiculturalism and postcolonialism as both cultural facts and critical concepts are important for understanding how Australia has represented itself at Eurovision. As Sneja Gunew (2004) has suggested, multiculturalism and postcolonialism are "adjacent to and overlapping" critical theories that are integral to understanding how race, ethnicity, and nation have been constructed over time in the Australian context, and how Australia might relate culturally and politically to other parts of the world. Certainly, in much discourse surrounding participation in Eurovision, including the Eurostralia skit, Australia is presented as an extension of Europe, but this extension is not simply in the sense of an *addition to* but also an *expansion upon*. Australia is undeniably a product of European expansionism via both colonisation and migration, but it is not solely the product of Europe. It is shaped also by its Indigenous heritages, including the fraught relations that attend its colonial legacies, and its geographical location within the Asia-Pacific. This constellation of factors is emphasised through the use of visibly different and specifically Asian Australians as part of the Australian Eurovision delegations and representatives. Mauboy, Sebastian, Im, Pang, and Chin, each with their own Asian heritage, personify the other cultural and geopolitical forces upon the development of Australia as a Western society somewhat paradoxically situated in Asia, acknowledging, of course, that this is a category as problematic and as elastic as "Europe" itself. Australia has sent as its representatives not just those who represent the European connection, such as French Australian Zemiro and Portuguese Australian Gonçalves, but also those who represent the complexities of Australia's Indigenous and migrant heritages, such as Mauboy and Firebrace, and its geopolitical paradox. In a utopic interpretation, this performance of multicultural, geopolitically complex and playful Australia presents a model for the "unity in diversity" desired by Eurovision and the European community more broadly.

References

Adams, Cameron. 2015. "Guy Sebastian Isn't the First Australian to Sing at Eurovision—Here's the Others." *News.Com.Au*, May 22. Accessed January 29, 2016. http://www.news.com.au/entertainment/music/guy-sebastian-isnt-the-rst-australian-to-sing-at-eurovision–heres-the-others/news-story/3a1cd7a783fe4c6168a88e5bfc04661f.

Balibar, Etienne. 2009. "Ideas of Europe: Civilization and Constitution." *Iris* 1 (1): 3–17.

degli Alessandrini, Claudia. 2015. "Eurovision Song Contest: A Platform for European Identity Contestations and Imaginary Membership Narratives." MA thesis, Universiteit Van Amsterdam, Amsterdam.

Edwards, Ken. 2009. "Traditional Games of a Timeless Land: Play Cultures in Aboriginal and Torres Strait Islander Communities." *Australian Aboriginal Studies* 2: 32–43.

Eurovision Song Contest. 2011. *Lordi—Hard Rock Hallelujah (Finland) 2006 Eurovision Song Contest Winner*. YouTube. Accessed January 29, 2016. https://www.youtube.com/watch?v=gAh9NRGNhUU.

Eurovision Song Contest. 2012. *Loreen—Euphoria—Live—Grand Final—2012 Eurovision Song Contest.* YouTube. Accessed January 29, 2016. https://www.youtube.com/watch?v=Pfo-8z86x80.

Eurovision Song Contest. 2014. *Eurovision Song Contest: Down Under (Interval Act by Australia; Jessica Mauboy).* YouTube. Accessed January 29, 2016. https://www.youtube.com/watch?v=tUsJi8nKsj0.

Eurovision Song Contest. 2015. *Guy Sebastian—Tonight Again (Australia)—LIVE at Eurovision 2015 Grand Final.* YouTube. Accessed January 29, 2016. https://www.youtube.com/watch?v=H0EhhZWXTng.

Faruqi, Osman. 2017. "Cook and Sing for Us, but Don't Date Us: What Reality TV Tells Us About Australia and Diversity." *Junkee*, July 26. http://junkee.com/diversity-reality-tv-australia-bachelor/114455.

Gray, Stephen. 2015. "The Northern Territory Intervention: An Evaluation." Melbourne: Castan Centre for Human Rights Law, Monash University.

Gunew, Sneja. 2004. *Haunted Nations: The Colonial Dimensions of Multiculturalisms.* London: Routledge.

Guy, Stephanie B. 2015. "Bodies, Myth and Music: How Contemporary Indigenous Musicians Are Contesting a Mythologized Australian Nationalism." *ESharp* 23 (Spring): 1–21.

Hage, Ghassan. 1998. *White Nation: Fantasies of White Supremacy in a Multicultural Society.* Sydney: Pluto.

Hegarty, Siobhan. 2015. "7 Reasons Why Guy Sebastian Was Born a Eurovision Star." *SBS.* Accessed January 29, 2016. http://www.sbs.com.au/programs/eurovision/article/2015/03/04/7-reasons-why-guy-sebastian-was-born-eurovision-star.

Huijser, Hendrik. 2007. "Australian Idol Versus Cronulla: Whither the Postcolonising Nation?" *New Zealand Journal of Media Studies* 10 (2): 131–143.

Idato, Michael. 2018. "Jessica Mauboy Locks in We Got Love as Australia's Song for Eurovision." *Sydney Morning Herald*, March 6. https://www.smh.com.au/entertainment/music/jessica-mauboy-locks-in-we-got-love-as-australia-s-song-for-eurovision-20180306-p4z350.html.

Ismayilov, Murad. 2012. "State, Identity, and the Politics of Music: Eurovision and Nation-Building in Azerbaijan." *Nationalities Papers* 40 (6): 833–851.

Jordan, Paul. 2014. *The Modern Fairy Tale: Nation Branding, National Identity and the Eurovision Song Contest in Estonia.* Tartu: University of Tartu Press.

Lam, Celia. 2018. "Representing (Real) Australia: Australia's Eurovision Entrants, Diversity and Australian Identity." *Celebrity Studies* 9 (1): 117–125.

McIntyre, Joanna. 2017. "Transgender Idol: Queer Subjectivities and Australian Reality TV." *European Journal of Cultural Studies* 20 (1): 87–103.

Molloy, Shannon. 2016. "It Costs a Fortune to Compete and We're 14,000 km Away … So Why Do We Bother with Eurovision?" *News.com.au.* https://www.news.com.au/entertainment/tv/it-costs-a-fortune-to-compete-and-were-14000km-away–so-why-do-we-bother-with-eurovision/news-story/b53407a6444fe03f423ec38e2d1b3ea3.

Mutsaers, Lutgard. 2007. "Fernando, Filippo, and Milly: Bringing Blackness to the Eurovision Stage." In *A Song for Europe: Popular Music and Politics in the Eurovision Song Contest*, edited by Ivan Raykoff and Robert Deam Tobin, 61–70. Aldershot: Ashgate.

News.com.au. 2016. "Olivia Newton-John Almost Didn't Play Sandy in *Grease*." January 26. Accessed January 29, 2016. http://www.news.com.au/entertainment/movies/olivia-newtonjohn-almost-didnt-play-sandy-in-grease/news-story/6488b5668dfb5fe97136eeb560ba9e48.

Pagden, Anthony. 2002. *The Idea of Europe from Antiquity to the European Union*. Woodrow Wilson Center Series. Washington, DC: Woodrow Wilson Center Press.

Pardy, Maree, and Julian C. H. Lee. 2011. "Using Buzzwords of Belonging: Everyday Multiculturalism and Social Capital in Australia." *Journal of Australian Studies* 35 (3): 297–316.

Ramm, Jennifer. 1989. "Learning the 'Australian Myth': An Analysis of Australian Folk Songs." *Journal of Australian Studies* 13 (25): 23–33.

Roxburgh, Gordon, and Paul Jordan. 2017. "Shining a Light on the United Kingdom: 60 Years at Eurovision." *Eurovision.tv*, January 12. https://eurovision.tv/story/shining-a-light-on-the-united-kingdom-60-years-at-eurovision.

Scott, Rosie, Anita Heiss, and Brenda L. Croft, eds. 2015. *The Intervention: An Anthology*. Sydney: Concerned Australians.

Screen Australia. 2016. *Seeing Ourselves Reflections on Diversity in Australian TV Drama*. Sydney: Screen Australia.

Sieg, Katrin. 2013. "Conundrums of Post-socialist Belonging at the Eurovision Song Contest." In *Performing the 'New' Europe: Identities, Feelings and Politics in the Eurovision Song Contest*, edited by Karen Fricker and Milija Gluhovic, 218–237. Houndmills: Palgrave Macmillan.

Tan, Monica. 2016. "Graham Norton Is Right, Australia's Participation in Eurovision Is Stupid." *Guardian*, May 11. https://www.theguardian.com/tv-and-radio/2016/may/11/graham-norton-is-right-australias-participation-in-eurovision-is-stupid.

Van Ee, Dennis. 2017. "Australian Delegation: We Need to Be the Sweden of the South to Stay in Eurovision." *ESC Daily*. https://www.escdaily.com/australian-delegation-need-sweden-south-stay-eurovision/#.

Vuletic, Dean. 2018. *Postwar Europe and the Eurovision Song Contest*. London: Bloomsbury.

Werbner, Pnina. 2013. "Everyday Multiculturalism: Theorising the Difference Between 'Intersectionality' and 'Multiple Identities'." *Ethnicities* 13 (4): 401–419.

West, Chris. 2017. *Eurovision: A History of Modern Europe Through the World's Greatest Song Contest*. London: Melville House.

Wise, Amanda, and Selvaraj Velayutham, eds. 2009. *Everyday Multiculturalism*. Basingstoke: Palgrave Macmillan.

CHAPTER 7

"Every Region of the World Should Have a -Vision": Eurovision in the Asian Century

Abstract In 2016, it was announced that the EBU had partnered with SBS to establish a new song contest for the Asia-Pacific region, but fans remain sceptical of the project's viability. One challenge is the existence of other song contests in the region established by the Asian-Pacific Broadcasting Union. Other challenges emerge from regional politics and Australia's own shifting relations within the region. This chapter explores fan perceptions of Australian foreign relations through the lens of the Eurovision Song Contest and the proposed Eurovision Asia Song Contest. Within this, it also considers other global forces, such as relations with the United States, Americanisation, and globalisation.

Keywords Foreign relations · Asia-Pacific region · Americanisation · Globalisation · Cultural relations

Australia's entrance into the Eurovision Song Contest heralds a new era for the contest as it continues to expand. As executive supervisor Jon Ola Sand stated in his announcement that Australia would return again in 2016, "We strongly believe the Eurovision Song Contest has the potential to evolve organically into a truly global event. Australia's continued participation is an exciting step in that direction" (quoted in BBC 2015). It was subsequently announced that the EBU and SBS

J. Carniel, *Understanding the Eurovision Song Contest in Multicultural Australia*,
https://doi.org/10.1007/978-3-030-02315-7_7

would be collaborating on the launch of a new song contest for the Asia-Pacific region, Eurovision Asia. Ola Sand highlighted the development as an opportunity for Asian artists to "perform on the Eurovision stage", while EBU Reference Group Chair Dr. Frank-Dieter Freiling spoke of Eurovision's appeal as "crossing continental borders" (SBS 2016). Then SBS Managing Director Michael Ebeid's responding statement articulates the complex geopolitical and cultural status of contemporary Australia while also claiming authority over the event:

> As the official broadcaster of Eurovision for over 30 years, SBS are pleased to explore the opportunity to bring an event of this calibre more closely to our shores, strengthening the multicultural ties in our own backyard. Asia Pacific has a spectacular music culture and the perfect next step to extend the Eurovision brand, bringing its hugely popular appeal beyond Australian audiences and to the wider region. (SBS 2016)

Eurovision Asia is not the first time the EBU has attempted to capture the Asian market via its song contest format. In 2008, the EBU unsuccessfully attempted to broker a deal with the Asia-Pacific Broadcasting Union (ABU), the regional equivalent to the EBU, but later sold the rights to the "-vision" brand and song contest format to a Singaporean company. The contest name was later changed to "Our Sound: The Asia-Pacific Song Contest," although it still used Eurovision's signature heart logo within its branding. This change of name was executed at the request of the ABU who, like the EBU, use the "Asiavision" brand for their news exchange service; it is also for this reason that the new proposed contest will be called Eurovision Asia, although many fans still refer to it colloquially as "Asiavision." The Our Sound contest never launched, and in 2010 was postponed indefinitely. In 2011, the ABU announced it would begin its own song contest; this was as much a revival of their own ABU Popular Song Contest that ran from 1985 to 1987, and its ABU Golden Kite Song Festival, which, in addition to Asia-Pacific nations, had included Finland and former Yugoslavia, as it was based upon the Eurovision Song Contest.

The ABU's twin festivals, the Radio Song Festival and the TV Song Festival, launched in 2012. Initially biennial, the Radio Song Festival has been annual since 2014, with the exception of 2017, when the host country Thailand cancelled the event due to the death of King Bhumibol Adulyadej. The TV Song Festival is an annual event featuring signed or established artists submitted to represent their nation by television broadcaster members of the ABU. The Radio Song Festival most resembles Eurovision as it is a competition between national representatives, while the TV Song Festival is a non-competitive gala event, but unlike Eurovision, the ABU Radio Song Festival is aimed at unsigned artists who are submitted by radio rather than television broadcasters and it is not limited to public broadcasters; that is, no single broadcaster has the responsibility of submitting a national entry. Consequently, a nation may have more than one representative as these can be submitted by different broadcasters in the same country. Furthermore, the contest is assessed only by an international jury comprised of representatives from various ABU media organisations and it is not broadcast live; participating countries agree to broadcast the festival within a month of the live event. Although the ABU serves the Asia-Pacific region and its initial song festivals focused on artists from that region exclusively, in 2016 Romania joined the competition.

The first festival initially attracted twenty-six entries from eighteen radio broadcasters or industry bodies (Australia's entry, Danielle Blakey, was submitted by Commercial Radio Australia) from fifteen nations; fifteen were selected for the grand final, but only thirteen performed as Fiji and Sri Lanka withdrew from the competition for unknown reasons. South Korea won the Grand Prix with duo Billy Acoustie's "For a Rest"; notably, Australia won the Gold Prize (second place) with Danielle Blakey's "Fearless" and is credited also with the Bronze Prize (fourth place), although its winner Sammy Ray Jones, who is Indigenous Australian, was technically representing Vanuatu with the support of the ABC. The format of the festival changed from this competitive format after the inaugural event; while entries still qualify for a "final," prizes are no longer awarded. Consequently, there is no "winning" nation to host the next festival, so its rotation is determined by the ABU; this was also the practice even following South Korea's win in the competitive year. Australia was slated to host in 2014, but for unknown reasons this did not occur and Sri Lanka stepped into the role instead.

Australia's participation in the Radio Song Festival has been limited, as has its involvement in the TV Song Festival. Commercial Radio Australia submitted entries to represent Australia in the 2012 and 2014 Radio Song Festivals, but withdrew from the festival in 2015 and has not submitted an entry since. Australian artists have participated in the TV Song Festival three times, from 2012 to 2014 but, as with the

Radio Song Festival, Australia withdrew in 2015. Notably, Dami Im was Australia's representative in 2014, two years prior to her success at Eurovision. She was preceded by Havana Brown in 2012 and Justice Crew in 2013. As SBS is also involved in process for the TV Song Festival as well as Eurovision, it is feasible that the broadcaster has opted to invest more heavily in the event with which it shares a longer standing tradition and from which it might develop its own profitable brand.

Official press releases and much media coverage for the proposed Eurovision Asia Song Contest neatly overlook the preceding history of Our Sound and the existence of the ABU song festivals. Certainly, there are differences between the contest and the festivals, such as the contest's focus on a competitive, live format rather than a non-competitive event that is recorded for a delayed broadcast at the individual broadcasters' discretion. Nevertheless, the proposed contest will undoubtedly be in competition with the festivals, and SBS will require the cooperation of many national broadcasters who already participate in the TV Song Festival. This adds an extra element of industry politics to the already complex politics of Australia within the Asia-Pacific region.

Historically, Australia's relationship with the Asia-Pacific region, and the very concept of Asia itself, has been ambiguous. The settler society of Australia identified racially and culturally with its British colonisers, and through this Europeanness or Westernness more generally, and felt its distance from Europe and its difference from nearby Asia keenly as a result; this is the underlying logic of Geoffrey Blainey's (1991) tyranny of distance. This resulted in various spatial and racial anxieties about Asia played out in a range of policies in the nineteenth and twentieth centuries, usually centred upon immigration but tied also to labour and trade. Australian economic policy in the twentieth century was, however, more pragmatic in its approach to the Asia-Pacific region than the paranoid underpinnings of its immigration policy and in some ways assisted the slow erosion of the so-called White Australia policy. In terms of foreign policy and security, legacies of old World War II anxieties about the region remain in the ANZUS treaty with the United States. When he was federal treasurer, Paul Keating once referred to Asia as "the place you fly over to get to Europe" but as prime minister he was dogged in his pursuit of productive Asian relations (see Keating 2000). Deemed the "Asian century," twenty-first century foreign and economic policies are very much focused on the Asia-Pacific region, but with an awareness of how Australia can leverage its historical and cultural ties to Europe and the US. Australia's belonging to its geographical region is, however, conditional. It is defined by geography, experienced primarily through economy, and limited by race and culture. As former Singaporean Prime Minister Lee Kwan Yew once quipped, Asian countries would continue to build useful links with Australia but "we could never regard you as *family*" (quoted in Dalrymple 2003). Accordingly, various Australian cultural institutions, such as Australian involvement in the Asian Football Confederation and now the development of Eurovision Asia, play an increasingly important role in fostering stronger cultural and affective relations. As SBS's Ebeid stated with regard to Eurovision Asia, "SBS sees entertainment as one area in which we can play an active role in strengthening Australia's ties with Asia as part of our overall commitment to reflecting multiculturalism and contributing to the social health of our communities" (SBS 2016).

In his statement for the announcement of Eurovision Asia, Ola Sand's invocation of Eurovision's original aims in 1956 of "uniting nations and exploring new television technology" (SBS 2016) suggests similar goals for the new contest. The history of the Eurovision Song Contest, as discussed in Chapter 2, is one of pragmatism mixed with idealism. It was pragmatic in that it sought to share resources, monitor potential conflicts, and further develop television technologies, and it was idealistic in that it sought to create a unified European public through shared information and cultural productions, such as the Eurovision Song Contest. The latter element was also clearly a political project as it presumes shared values and ideologies; the Europe at its core was situated clearly on the Western side of the iron curtain. Decades after the dismantling of the USSR, old East/West politics still inform Eurovision. Situated in a vastly different historical and geopolitical context, Eurovision Asia could nevertheless replicate this ideological vision because of the centrality of Western values within this—and the difference between "Western" and "Asian" values has played a strong role in the marginalisation of Australia in Asia, underpinning Lee's comment about Australia never being regarded as "family" by nations in the region. While the official narrative of Eurovision Asia seeks to create connections within the region, it may be impeded by perceptions that it is a new form of cultural colonisation. As one participant ponders: "I guess it depends on how Asia wants itself to be seen - a replica of a Western model or something different?" Similarly, interviewee Daniel observes, "not everyone in Asia thinks we're Asia, so it could definitely come across as being imposed."

In imagining the contest's expansion as a general principle, fans express caution around rendering Eurovision vulnerable to the slippery slope of globalisation or, more specifically, Americanisation. For some, Australian involvement and its role in the proposed expansion into Asia positions Australia as an agent of this homogenising process because of its close connection to American popular culture. "European music and performers are different," one participant explains of their view on Australian participation, "and I don't want it turning into one bland Americanised show." Yet the expansion into Asia in general, even without Australian involvement, is equally seen as a potential homogenisation of the contest. Discussing their perspective on Eurovision Asia, another participant states, "I just envision a lot of homogenous K-pop type music, boring."

Fear of the Americanisation of Eurovision belies the fear of Australia's own vulnerability to Americanisation. Australia shares many political and ideological affiliations with the US, to the extent that some voice concerns about the nation's role as the "fifty-first state" (see, for example, Altman 2006; Pilger 2007), and Australians are, like much of the world, avid consumers of American popular culture. Yet, as David Carter and Bridget Griffen-Foley emphasise, "The dynamics of contemporary culture involve Australians both participating in and resisting global cultures;" the Eurovision Song Contest perhaps exemplifies this tension. Eurovision does screen in the United States, enjoying moderate success for the past few years on Logo, a cable channel known for its queer-oriented programming, but fans are sceptical of America's ability to "get it." Each year, online articles seeking to explain the contest to Americans proliferate in response to social media posts about Eurovision from bemused Americans, or media content such as "Americans react" videos from various sources on YouTube. Similar concerns are also reflected in reactions to the announcement that American comic actor Will Ferrell will be creating a Eurovision comedy for international streaming service Netflix (Stephens 2018). Although Ferrell claims an extended familiarity with the contest via his Swedish wife, fans are concerned that the film will play too heavily upon stereotypes of quaint European backwardness and substandard pop music, which often emerge from the "Americans react" videos.

Australia's longstanding fandom of Eurovision—and fans' desire to preserve the contest's authenticity by excluding the global cultural powerhouse of the US—is offered up by some as evidence that Australia remains distinct from the United States in cultural tastes and attitudes.

While appreciation of and participation in Eurovision does not completely dismantle Australian concerns about Americanisation, nor does it necessarily signal a renewed "Europeanisation" of Australia, it does help articulate Australia's hybrid political and cultural scene. Fans both for and against Australian participation are particularly wary of the prospect of American involvement for both cultural and political reasons, and are reluctant for Australia to be the gateway for this development: "we don't want to be the reason they use to try to justify bringing in America." This is not simply a matter of fear of American cultural imperialism and domination of the global music industry, but also a reluctance to over-politicise the contest: "opening [Eurovision] up to other nations such as the USA and China would make it seem like a battle of the superpowers rather than just a genuine contest."

As discussed in the previous chapter, opinions on Australian involvement in Eurovision are divided. They are also divided on the question of Eurovision Asia, but generally the prospect is viewed with optimistic curiosity—alongside a healthy dose of scepticism as to whether the contest will ever be launched, as it was originally scheduled to commence in 2017 but this did not eventuate. Many participants see it as an appropriate explanation for ongoing associations between Australia and the song contest. Australia's participation in Eurovision is viewed by some as a practical strategy to legitimise the proposed Asia-Pacific chapter of the contest, so it is "within the EBU's interest to keep us in until our own competition launches." For some, the launch of Eurovision Asia should be the limit of participation—"it potentially means that Australia can stop this charade of Eurovision participation"—while others lament that it will lead to Australia's exclusion from the original contest. "I would be devastated if Asiavision [sic] meant Australia could no longer participate in Eurovision," reflected one participant, while another observed, "This seems like a way of kicking Australia out of EUROvision [sic]. Not overly inclusive in my book."

Australian fans on both sides of the question of Australian involvement in the original Eurovision Song Contest express nuanced views about geopolitics and globalisation as part of their rationale, although opponents do generally cite geography as their primary objection. One response—"Europe is a part of us. We are not a part of Europe"—expresses the complexity of identity in contrast with geography. For some, perceived European antipathy about Australian involvement needs to be respected in order to preserve positive cultural relations with that region, so Australia ought to depart from the original contest while goodwill remains. The pragmatics and fairness of the contest's administration is a significant concern for fans, with some suggesting that Australia's involvement "tak[es] up a qualifying spot for countries which are in Europe and have been trying for years." Proponents of Australian involvement provide an economic counter to this point, arguing that Australian involvement opens rather than closes opportunities. By this reasoning, Australia's involvement facilitates increased participation as "the money that we inject into the contest allows them to make it cheaper for other countries that might be struggling to afford participation." In the eyes of these fans, Australia's economic contribution is therefore important to the contest's perpetuity at a time when economic instability leads many nations to drop out of the contest due to economic reasons. This is an interesting point to consider in relation to concerns that Australian participation is ruining a "European tradition," as it suggests that Australia helps to preserve it economically, even as it might dilute it culturally and symbolically. For others, the concern is centred more upon the effect of involvement in an ostensibly European tradition on Australian relations within Asia as it may exacerbate reservations about Australia's western Otherness within the region. If Eurovision participation signaled legitimacy as a "European" country for Israel (Lemish 2004, 42–43) and Azerbaijan (Carniel 2015), for Australia it signals belonging to a Eurocentric Western culture and identity that is neither contained nor limited by geographical boundaries. This has significant implications for Australian relations within Asia, where it is still seen as the outsider, in part because of its ideological connection to the West and its role in British colonialism. As one participant observed, by focusing on Europe via Eurovision "Australia seems like it's shunning Asia," which they saw as illogical and "messy" for both Australian and European geopolitics, as well as the administration of the contest itself.

Within their responses, participants demonstrate complex ideas about Australian, Asian, and European culture and politics. These frequently centre upon perceived ideological differences between these sites. From this emerges an image of "Asia" as divided, totalitarian, and too serious to embrace the kitsch irony with which the original contest is imbued, in contrast to a generalised perspective of "fun" Western liberalism in Australia and Europe: "the culture here [in Asia] is different. In Europe it's more fun and winning doesn't matter (except for Sweden lol) but in Asia I think it will be more serious with countries actually wanting to win." Diversity within the region, one of the core characteristics of the original contest valued by its Australian fans, is perceived by many as one of the greatest challenges Eurovision Asia faces. One participant observes that the "Asia-Pacific is even more varied than Greater Europe," but is hopeful that "this contest would break cultural barriers." Another is less optimistic: "Too many cultural differences for a cohesive vote?" While China is identified by some as a specific "road-block" to Eurovision Asia's success, many view the region as too politically fragmented to be able to collaborate for the contest's success. Ebeid agrees; upon his October 2018 departure from the Managing Director role of SBS, he nominated the delays in launching the contest as his greatest disappointment, describing the project as "too geopolitically difficult" (quoted in Knox 2018). Within the original contest, the politics, while sometimes distasteful and disruptive, are accepted by most Eurofans as simply part of the parcel: "When it's about countries competing, politics will always come into it." Alongside this realist perspective of the contest's politics, fans are nonetheless heavily invested in the idea of Eurovision as an event that generally succeeds in getting Europe to set aside their differences for a night of music and unity. Asian politics, on the other hand, are perceived as too much of a challenge: "the contest would become much more heavily politicised than Eurovision, as there is a lot more political tension in Asia, including many more dictatorships." Only a handful of participants identified the possibility that Eurovision Asia could perform the same function for a fragmented or divided Asia as it arguably performs for Europe: "Eurovision was founded to bring Europe together after WW2 [sic], and it's been largely successful in that mission. Asia has remained bitterly divided since that time, and it's long past time that similar efforts were made to bring the region together. I think Asiavision [sic] could be on such mechanism to make that a reality."

Undoubtedly, the prospect of being at the ground level of an event designed to unite the Asia-Pacific region is politically and economically appealing for Australian policy-makers, as well as to the Australian television and music industries. In the fragmented image of Asia reflected by participant responses, we also get the persistent sense that Australia does not quite belong in Asia—but nor does it necessarily belong in a European event. In its very branding, Eurovision Asia articulates the ongoing tension Australia experiences between its history and its geography. It is a hybridisation of its colonial and European migrant past, and its geopolitical future. Slated to be the first host for the proposed event, Australia is well positioned to build a bridge, to draw upon the theme of its first Eurovision as a participant in 2015, between the Asia-Pacific and Europe, while also fostering stronger cultural ties to its own geographical region.

References

Altman, Dennis. 2006. *51st State?* Melbourne: Scribe.

BBC. 2015. "Australia to Return to Eurovision Song Contest in 2016." *BBC. Com*, November 17. Accessed January 29, 2016. http://www.bbc.com/news/world-australia-34841066.

Blainey, Geoffrey. 1991. *The Tyranny of Distance: How Distance Shaped Australia's History*. Rev. ed. Sydney: Pan Macmillan.

Carniel, Jessica. 2015. "Skirting the Issue: Finding Queer and Geopolitical Belonging at the Eurovision Song Contest." *Contemporary Southeastern Europe* 2 (1): 136–154.

Carter, David, and Bridget Griffen-Foley. 2013. "Culture and Media." In *The Commonwealth of Australia*, edited by Alison Bashford and Stuart Macintyre, 237–262. New York, NY: Cambridge University Press.

Dalrymple, Rawdon. 2003. *Continental Drift: Australia's Search for a Regional Identity*. Aldershot: Ashgate.

Keating, Paul. 2000. *Engagement: Australia Faces the Asia Pacific*. Sydney: Pan Macmillan.

Knox, David. 2018. "Australia Returning For Eurovision 2019." *TV Tonight*, October 2. Accessed October 2, 2018. https://tvtonight.com.au/2018/10/australia-returning-for-eurovision-2019.html.

Lemish, Dafna. 2004. "'My Kind of Campfire': The Eurovision Song Contest and Israeli Gay Men." *Popular Communication* 2 (1): 41–63.

Pilger, John. 2007. "Australia: The New 51st State." *New Statesman*, 24.

SBS. 2016. "SBS Secures an Exclusive Option to Develop the Eurovision Song Contest Concept in Asia." *SBS*, March 21. https://www.sbs.com.au/programs/eurovision/article/2016/03/21/sbs-secures-exclusive-option-develop-eurovision-song-contest-concept-asia.

Stephens, Heidi. 2018. "Eurovision Has Come Too Far to Be Made a Mockery of by Will Ferrell." *Guardian*, June 21. https://www.theguardian.com/commentisfree/2018/jun/21/eurovision-will-ferrell-film-song-contest.

CHAPTER 8

Epilogue

Abstract Written while attending the Eurovision Song Contest in Lisbon 2018, this final chapter reflects upon how the 2018 contest illustrated and challenged many of the trends and themes explored throughout the book. It concludes that Australia's middle-of-the-table result in 2018 should not be seen as a sign to leave the contest but a shift in its status as a competing nation, and an additional impetus in its project of establishing a new song contest for the Asia-Pacific region while still maintaining its connections to the original event and, through this, Europe.

Keywords Multiculturalism · Tourism · Live events · Cultural relations

The evening light of late spring is only just beginning to dim as the strains of "*Te Deum*" fill the air and the crowd cheers in response. A crowd of thousands is gathered in Lisbon's Praça do Comércio where the Eurovision Village has been set up for the duration of the 2018 contest. We have all been either unlucky or uninterested in getting tickets to the live grand final being filmed several kilometres away at Altice Arena. This crowd doesn't seem to care or notice that they are watching the contest on big screens in an open space rather than in the arena. They cheer, clap, and sing along; just as the arena audiences had for

J. Carniel, *Understanding the Eurovision Song Contest in Multicultural Australia*,
https://doi.org/10.1007/978-3-030-02315-7_8

the semi-finals I attended days earlier. Where I am standing, I am surrounded by Spanish, French, and UK supporters. Ahead of me in the crowd I can see Ukrainian, Irish, and Israeli flags. Looking around I can see clumps of Australian fans waving flags and inflatable boxing kangaroos. One fan is in full drag with a pink wig like a cloud of fairy floss on their head, while another group sports rainbow angel wings. We are all huddled for warmth against the spring air blowing cold off the Tagus River. I have been standing here for just over two hours, unwilling to move in case I lose my hard earned spot with a good view of the large screen on the stage ahead. I have a strange nostalgia for my blanket and couch, and my cup of tea after the 5 a.m. alarm.

Attending the Eurovision Song Contest in the host city for the first time is certainly the pinnacle of my own fandom and my fieldwork. Although I was disappointed to be unsuccessful in getting tickets to the actual grand final show, this gathering in a public space feels more like the Eurovision I have known throughout my life than the live arena shows I had attended during the week. The Eurovision Song Contest is, after all, a television event and here I am watching it on a large screen with a makeshift Eurovision family. Where watching the filming had given me an invaluable and fascinating behind the scenes perspective, this public transmission allowed me to see the final result—the postcards, the graphics, the staging designed for me as a television viewer not as a member of a live arena audience. Instead, I am at one of the biggest Eurovision parties in town, possibly the biggest in the world, even as others gather at the sites where the voting live crosses will occur.

Australia's voting ambassador in 2018 is Portuguese Australian and openly gay newsreader—and lifelong Eurovision fan—Ricardo Gonçalves, controversially replacing Lee Lin Chin. Although he delivers the votes in English, Gonçalves takes a moment to greet his relatives in Madeira in Portuguese. It is a moment that makes the news in Portugal, fitting well with the sea-faring theme of "All Aboard" that sought to emphasise Portuguese culture as mobile and global by showcasing other Lusitanic cultures and musical traditions. For Australian audiences, Gonçalves' family greeting reasserts belonging in the contest by reminding viewers of the real, personal, and affective attachments Australia and Australians have to Europe through migration. Gonçalves' sexuality is not foregrounded in the same way but is nevertheless significant to his representing the Australian votes for the Eurovision Song Contest, an event with strong ties to the LGBTI community.

While Gonçalves constituted an attempt to connect meaningfully with the host country, I witnessed a bizarre counter to this while watching a segment being filmed for SBS's *The Feed*. Australian fans were called via social media to gather in an Irish pub in Lisbon. Cramped in a corner decorated with flags and behind a table covered in glasses of beer, a bit was played out between host Jan Fran and an Australian fan, Nelson. Asked to reveal his "true colours", Nelson took off his Australian t-shirt to reveal a Portuguese t-shirt. "You've got both Portuguese and Australian citizenship," Fran demanded, "so where are your allegiances? Where do they lie right now?" In the clip that aired, Nelson laughingly points to the room of Australians and says his allegiances are clearly with Australia right now (SBS 2018). The scene was filmed various times with different groups of fans in the reaction shots. In the take of this scene that I witnessed, the crowd booed Nelson until he sheepishly waved an Australian flag. It was an interesting manifestation of the concerns voiced by fans in my study that Australian participation lent itself too easily to a parochial nationalism that could undermine the multiculturalist ethos the contest represents for many.

Australian nationalism received a blow this year. Despite presenting with a solid song and the return of Jessica Mauboy, a familiar and well-admired face for Eurovision fans, Australia performed poorly relative to its past results. The crowd at the *praça* gasps and grumbles when Australia's public voting results are revealed as the lowest of all entrants. Mauboy's "We Got Love" had premiered strongly in bookmakers' predictions. In the lead up to the contest, the feeling amongst Australian fans had been that if anyone could win the Eurovision Song Contest for Australia, it had to be Jessica Mauboy. The song played on the car radio as I drove to the airport to catch my flight to Europe—surely this was a sign? After its strong start, "We Got Love" rapidly slid to the middle of the odds as competition between the two pop bangers from Israel and Cyprus grew fierce and other sleeper hits gained greater traction amongst fans. At the contest, the goodwill toward Australians and the Australian entry has been palpable. "We love Jess!", my AirBnB host gushed at me as I checked in at the beginning of my trip. Later, as I checked out, he commiserated with me, unable to believe the results. Back home, various cultural and political commentators began to murmur: has Australia outstayed its welcome at Eurovision (ABC 2018; Carmody 2018; Lo 2018; Holden 2018)? Opinions on the initial and ongoing inclusion of Australia in the contest have always been divided in the fandom, and will remain so for the foreseeable future, but it is an exaggeration to see it as a decline in Australian-European cultural relations. Nevertheless, Australia's lacklustre results in 2018 might be the impetus to see Eurovision Asia finally launched.

Perhaps the future of Australian participation is uncertain, or perhaps competition was tough this year. The ebbs and flows of favour with the European public are all part of the experience in participating in the contest. Australia has now experienced the exhilarating heights of Dami Im's near win and the devastating low of Jessica Mauboy's loss. Fair-weather viewers might be deterred, but true Australian fans of the Eurovision Song Contest will adhere to the message of Mauboy's (Egizii et al. 2018) song to not give up, "'Cause we got love."

References

ABC. 2018. "Israel Celebrates Eurovision Win." *RN Breakfast*. ABC Radio National, May 14. http://www.abc.net.au/radionational/programs/breakfast/israel-celebrates-eurovision-win/9757590.

Carmody, Broede. 2018. "'Terrible': Eurovision Historian Criticises Jessica Mauboy." *Canberra Times*, May 14. https://www.canberratimes.com.au/entertainment/music/terrible-eurovision-historian-criticises-jessica-mauboy-20180514-p4zf42.html.

Egizii, Anthony, David Musumeci, and Jessica Mauboy. 2018. *We Got Love*. Sony.

Holden, Marr. 2018. "Eurovision 2018: The Joke Is Finally Wearing a Bit Thin." *Sydney Morning Herald*, May 14. https://www.smh.com.au/entertainment/celebrity/eurovision-2018-the-joke-is-finally-wearing-a-bit-thin-20180514-p4zf68.html.

Lo, Jieh-Yung. 2018. "The Novelty Has Worn Off—Let's Ditch Eurovision for Asiavision." *Sydney Morning Herald*, May 14. https://www.smh.com.au/entertainment/music/the-novelty-has-worn-off-let-s-ditch-eurovision-for-asiavision-20180514-p4zf53.html.

SBS. 2018. "Jan Fran Meets the Biggest Eurovision Fans." *The Feed*. https://www.sbs.com.au/programs/video/1228883011610/Jan-Fran-meets-the-biggest-Eurovision-fans.

Index

J. Carniel, *Understanding the Eurovision Song Contest in Multicultural Australia*,
https://doi.org/10.1007/978-3-030-02315-7

X

Z

The manufacturer's authorised representative in the EU is Springer Nature Customer Service Centre GmbH, Europaplatz 3, 69115 Heidelberg, Germany. If you have any concerns regarding our products, please contact ProductSafety@springernature.com

Printed and bound by CPI Group (UK) Ltd, Croydon, CR0 4YY
30/09/2026
02225666-0004